D1569569

SCHOOLCRAFT COLLEGE LIBRARY
3 3013 00027 7982

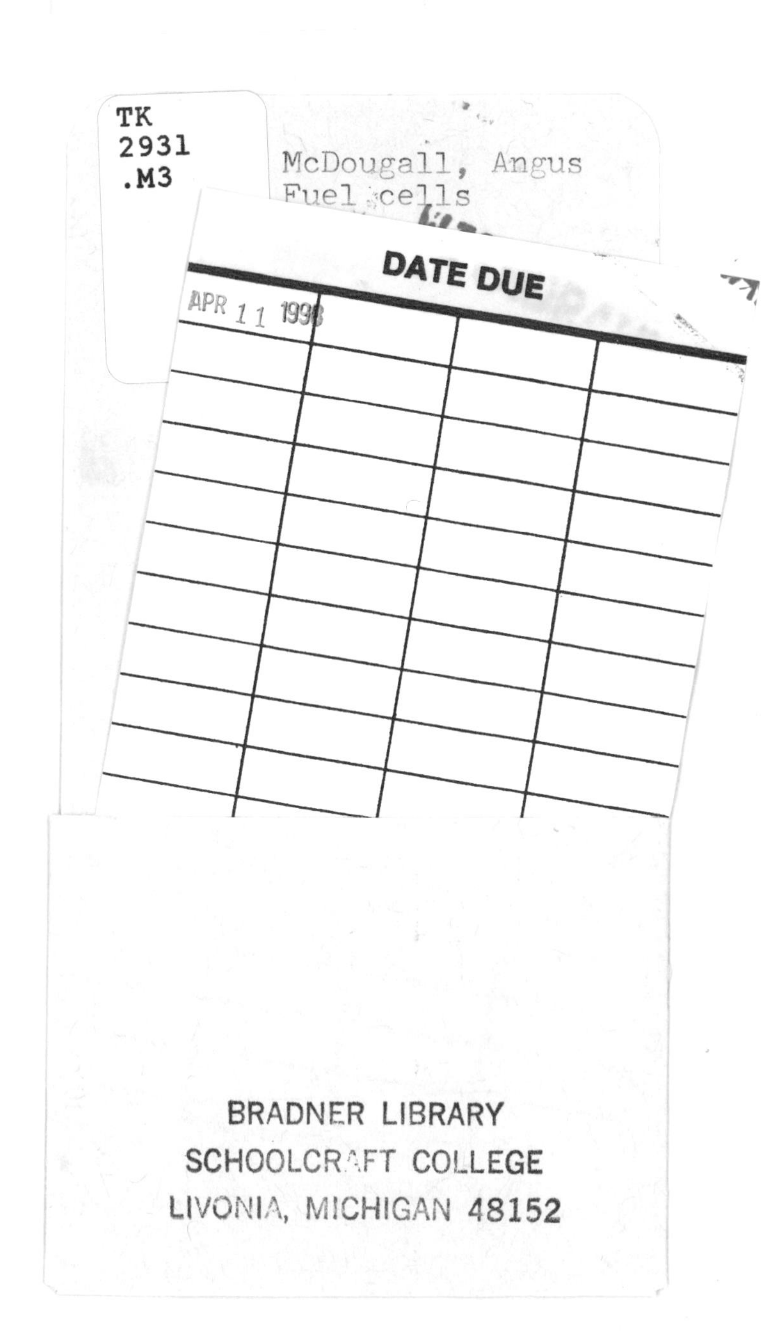
DATE DUE
APR 11 1998

FUEL CELLS

Also in the Energy Alternatives series:
HYDROGEN ENERGY
C. A. McAuliffe

ENERGY ALTERNATIVES SERIES
Series editor: C. A. McAuliffe,
Senior Lecturer in Chemistry, University of Manchester Institute of Science and Technology

FUEL CELLS

by

ANGUS McDOUGALL
Lecturer in Chemistry,
University of Manchester
Institute of Science and Technology,
Manchester, United Kingdom

A HALSTED PRESS BOOK

JOHN WILEY & SONS
New York

First published 1976 by
The Macmillan Press Ltd
London and Basingstoke

Published in the USA by
Halsted Press, a division of
John Wiley & Sons, Inc., New York

Printed in Great Britain

Library of Congress Cataloging in Publication Data

McDougall, Angus O 1934–
Fuel cells.

(Energy alternatives series)
"A Halsted Press book."
Bibliography: p.
Includes index.
1. Fuel cells. I. Title. II. Series.
TK2931.M3 621.35′9 76–15162
ISBN 0–470–15143–9

CONTENTS

EDITOR'S FOREWORD TO THE SERIES

The so-called 'energy crisis' is misnamed. There is certainly a serious problem regarding reserves of gaseous and liquid hydrocarbons, but immense fossil fuel reserves exist in the form of coal. If we take the United States of America alone and assume that only 50% of their known coal deposits are recoverable, then this would still be enough to last them for 2000 years at current rates of consumption. A similar situation applies in the Soviet Union.

However, crisis or not, the short oil embargo by the O.P.E.C. countries in 1973 did make us aware that we had not been giving due attention to our most important basic resource, energy. For example, the trend towards the use of naturally occurring hydrocarbons in most developed countries had been taking place without any thought of the long-term economic and social issues. It was also realised that although fossil fuels may be plentiful in the immediate future, complacency about energy is an irresponsible attitude and that we should be directing our efforts to seeking to use it more efficiently and to developing alternative sources.

Thus there has been in recent years a great deal of discussion and research into all aspects of energy sources and their utilisation. The questions considered range from the straightforward analysis of the efficiency of various energy-generating processes now in current use, through the possibilities raised by more unconventional sources such as solar energy and geothermal energy, to the demands which society has and will continue to make for greater supplies of energy to support itself in increasing comfort.

Within this range, it seems to me that there is a place for a series of books which sets down the essential scientific facts about the many different energy alternatives open to us once we look beyond current practice. With this in view, a number of experts have been approached and each asked to write a short account of his specialist subject and to outline its possible contribution to our future energy strategy. The choice of subjects must reflect to some extent my own prejudices about what may be considered important and relevant, but at the same time I believe that there is general agreement about what the major possibilities are. These will certainly be covered by the series. In scope each volume is principally restricted to the scientific and technological aspects of the subject, but it has occasionally proved impossible to avoid introducing broader social and political questions. This has not been discouraged, since discussion about the energy problem, although properly based upon a thorough appreciation of the factual background, must inevitably be conducted in a more general forum.

The series is intended to be open-ended, and thus will be better able to respond to changes in both technology itself and also in society's attitude towards it. It is not aimed at the specialist research worker—there are many books and articles which satisfy the needs of this market—but rather it is intended for university or polytechnic students, for engineers, planners, architects, policy-makers, and indeed anyone with a basic grounding in science who feels the need to be equipped with the facts before joining the debate about the future of energy.

Manchester, 1976 C. A. McAuliffe

FOREWORD TO THE BOOK

by

P. G. Ashmore (Professor of Physical Chemistry, University of Manchester Institute of Science and Technology)

Fuel cells seem to have been on the verge of capturing world markets as well as the imagination ever since my postwar years in Cambridge, when F. T. Bacon's researches were turning a backroom idea into a practical power unit. Admittedly, those were days when fuel was scarce in England, and electricity distribution networks were frequently overwhelmed by demand. Steady improvements in the efficiency of electricity production were achieved by bigger and better boilers, turbines and generators, and corresponding advances were made in the distribution system by reducing power losses, smoothing peak demands, and switching in and out parts of the national grid. In transport, it proved hard to find any competitor for the incredibly flexible and versatile petrol and diesel engines for road transport, and when diesel ousted steam locomotives it seemed that the internal combustion engine would be the dominant free power unit. Electrification of many intercity railways, however, has allowed the indirect methods of producing electricity to win back some business from diesel power. Can the direct production of electricity by fuel cells oust the internal combustion engine in any large market in the future?

In this book, a bold attempt is made to present the scientific and technical background to research and development of fuel cells, and to examine the economic and social factors that affect the use of fuel cells as primary producers of electricity in various parts of the world. It is surely a tribute to mankind's perpetual search for alternatives that research has gone as far as described into the properties of low-temperature and high-temperature fuel cells, into possible and practicable fuels, into improvements in efficiency and power-for-weight, into longer-lived and cheaper electrodes. All this work used fundamental researches on the kinetics and thermodynamics of electrode processes, and these are clearly outlined in the early chapters, at sufficient length to satisfy those with scientific training but not so deeply as to confound those whose science stopped when they left school. The last four chapters look at current practice, present costs, and possible future developments and uses of fuel cells. In presenting objectively 'the facts of the present and the aims of the future', this book certainly fulfils one of the main aims of the series.

Manchester, 1976 P. G. Ashmore

PREFACE

In the world of today the 'energy crisis' is frequently mentioned. It is widely assumed that we are being profligate with present-day sources of power and that eventually there will not be enough energy left in the world to provide for the predicted rapid increase in human population. In these circumstances it is not surprising that attention has been turned to ways of using our present resources of energy more efficiently, that is to say with less waste. The direct conversion of chemical energy into electricity by means of the fuel cell is an outstanding example of such a process, and much research effort has been expended in recent years in an attempt to develop the fuel cell for commercial use.

This book provides a discussion of the background to the use of the fuel cell as a source of electrical power, showing the advantages and constraints of the system and illustrating the attempts that have been made to produce commercially viable fuel batteries. Emphasis has been placed on general principles, which are discussed in the first four chapters, and on their application to particular designs of cell, described in the remainder of the book. Economic considerations are outlined briefly in chapter 12 and a review of future prospects appears as chapter 13. Although detailed descriptions, both of the scientific basis of fuel cell operation and of the technological development of fuel battery construction, have been reduced to a minimum, and there has been no attempt to describe the research methods used, references are made to suitable sources of further information on these points.

It is hoped that this book will provide a useful introduction to a subject of considerable interest today for a person having some knowledge of physics and chemistry but little other experience in the field. With a view to its ready acceptance in the modern educational environment, I have adopted, so far as possible, the methods of quantity calculus and the units of the Système Internationale throughout the book.

I should like to thank my colleagues and friends who have helped me during the writing of this book. I am particularly grateful for the encouragement, assistance and advice given by Professor P. G. Ashmore, Dr J. H. R. Clarke, Dr B. J. Tyler and Mr P. Wiseman, and especially Dr J. Lee who has read every word I have written and whose clear minded criticisms and suggestions I have rarely been able to ignore. I must also thank Chloride

Technical Ltd who have very generously provided much information, several of the diagrams and all of the photographs which appear in this book, and my typist, Mrs H. P. Robinson, who has coped splendidly with much unfamiliar scientific material.

Manchester, 1976 A. O. McDougall

CHAPTER 1

INTRODUCTION

That energy can neither be created nor destroyed is a very well-known statement of the Principle of Conservation of Energy, and we all accept its truth although certain processes taking place in nuclear reactions appear to confound it: thus when we speak loosely of 'energy production' we do not mean its production from nothing but the conversion of one form of energy into another form more useful for our purpose. It is this conversion which lies at the heart of human life on earth today, most of the characteristics of what we see as civilisation depending entirely on forms of energy conversion, primarily from the chemical energy of fossil fuels.

Direct and indirect energy conversion

Most of our available energy is obtained from the conversion processes

chemical energy → heat → electrical energy → mechanical work

and

chemical energy → heat → mechanical work.

These routes provide the basis for operation of the steam turbine electric power station and for the operation of the internal combustion engine. However, there have always been other methods of obtaining useful energy which do not involve heat as an intermediate. For example, the propulsion of boats (and occasionally land vehicles) by wind power has been well established for centuries, and windmills and windpumps also make use of the kinetic energy of air masses moving from high pressure to low pressure regions. In many countries production of electric power from the potential energy of water on its way to the sea is very important and has been established for many years. The idea of the direct conversion of chemical energy to electrical energy is not a new one either: it probably stems originally from the investigations of L. Galvani (1737–98) and A. Volta (1745–1827) who have both given their names to electrical phenomena. Both the primary Leclanché 'dry cell', powering torches and portable radios, and the secondary lead–acid cell, used for starting internal combustion engines, are familiar enough today, and there are now compact rechargeable alkaline cells which power such devices as portable electronic calculators. Here, the term 'secondary' is used to describe a cell which can 'store' electrical energy as chemical energy by a charging and discharging process, whereas a 'primary' cell is one which can only give electrical energy from chemical energy without the possibility of the reverse process occurring.

The conversion of mass to energy, as in the nuclear reaction, is not really an example of direct energy conversion since in all applications made so far the nuclear energy is immediately converted to heat which then produces electrical energy and work in the conventional way.

Since we are concerned with the conversion of latent chemical energy into a useful form from an undoubtedly dwindling supply which is moreover probably not capable of replacement, the efficiency of the conversion process is of paramount importance. By efficiency we mean the ratio of useful energy produced to the amount of energy supplied. The losses will be of energy in a non-useful form—for example, as waste heat, sound, or light. We shall see in detail in the next chapter how the conventional route from chemical energy via heat to electrical energy or mechanical work is subject to the restriction of a maximum efficiency which cannot be reduced by any technological improvement. This is the so-called Carnot cycle efficiency which restricts the operation of all heat engines. No such theoretical restriction of efficiency applies to systems of energy conversion operating without the intermediate production of heat, that is isothermal processes, but of course there are in both cases considerable practical restrictions of overall efficiency. It is likely that the larger the number of stages involved, the larger will be the consequential losses and the lower the overall efficiency. Once again, then, any system of direct conversion of chemical into electrical energy would seem to be advantageous.

As we have seen, such direct methods are—and have been—employed for many years commercially, but in rather restricted circumstances. They have acted as sources of electrical power in places where mains electricity generated by the conventional power station is impracticable, for example, in remote locations or in vehicles or for portable use, but the power produced by such installations is always very small compared with that from the steam turbine generator or motor generator system.

The chief reason for this small scale of operation is that the chemical energy used in the electrolytic cell must come from materials actually within the cell itself, and size and weight considerations are clearly a limitation. For a primary cell the chemical system must actually be built into the cell at the time of manufacture; for example, the dry cell used to power torches and radios must contain the zinc, magnesium dioxide, water and ammonium chloride necessary for the reaction

$$Zn + 2NH_4Cl + 2MnO_2 \rightarrow Zn(NH_3)_2Cl_2 + H_2O + Mn_2O_3$$

to proceed with the production of an electric potential and current. The reaction written here is undoubtedly not the complete reaction occurring within the cell and probably not the only source of the observed electric potential. However, the mechanism of operation of the Leclanché cell is even now not completely understood. The reader is referred to other works on electrochemistry mentioned in the Bibliography.

The lead–acid secondary cell (often called an 'accumulator') must contain enough lead sulphate and water to allow the application of an electric potential to drive the reaction

$$2PbSO_4 + 2H_2O \rightarrow Pb + PbO_2 + H_2SO_4$$

far enough to convert the required electrical energy into stored chemical energy. Thus 'charging' produces metallic lead and lead dioxide, which can then react in the presence of sulphuric acid to give electric power in the 'discharge' step. Lead is a very dense metal and the electrolyte, aqueous sulphuric acid, also has a high density so the weight of the unit involved is a clear limitation in the use of the lead–acid battery.

It is true that other primary and secondary cells have been devised which work on the same principle—that is containing all of the chemical materials required within the cell—but are less restricted in weight or size. The Ruben–Mallory cell, for example, uses zinc, mercury and potassium hydroxide to produce a compact primary cell of low mass suitable for hearing aids and miniature radio sets. The spontaneous chemical reaction yielding electrical energy is

$$Zn + HgO + H_2O \rightarrow Zn(OH)_2 + Hg$$

A secondary device which shows high promise of commercial viability is the sodium–sulphur battery where electrical energy is stored by producing molten sodium and sulphur from sodium sulphide. The charging reaction is

$$Na_2S(\ell) \rightarrow 2Na(\ell) + S(\ell)$$

and on discharge this reaction occurs in reverse.

Despite the obvious advantages of these two types of cell (and several others) they have not yet supplanted the Leclanché cell or the lead–acid battery in most of their applications. Both of these cells were first developed over 100 years ago (Leclanché, about 1866; the lead–acid battery by Planté in 1859) and have suffered comparatively few modifications in their long life. It is not easy to think of other inventions which have had such a long record of commercial success. The reason is undoubtedly that their competitors have usually been found to have other undesirable characteristics: for example, high capital cost (the Ruben–Mallory cell uses expensive materials) or difficult operating conditions (the sodium–sulphur cell must run at 250–350 °C to maintain the materials in the liquid state).

Fuel cells and related systems

Thus we may liken the currently used methods of direct conversion of chemical into electrical energy to a power station which can only operate with the amounts of coal and water actually on the premises when it is first started up—a situation in which it would clearly be impossible to generate electrical power for any length of time or at a high intensity.

There are two types of device which are more closely comparable to the power station and yet do involve the desirable direct conversion of chemical into electrical energy. One type comprises the various sorts of air depolarised cell, in which use is made of the ready supply of oxygen in the atmosphere to react electrochemically with a metal electrode. One example of this type utilises the reaction

$$Zn(s) + 4KOH(aq) + O_2(g) \rightarrow 2K_2ZnO_2(aq) + 2H_2O(\ell)$$

proceeding in the direction shown, and taking its supply of oxygen from outside the cell. The other type is the fuel cell where all the chemical materials concerned in the production of the chemical energy are outside the cell itself. It is the fuel cell with which this book is largely concerned (although some discussion of air depolarised cells will be found in chapter 9) since their uses may well overlap with those of true fuel cells and they have many of the same characteristics.

A fuel cell is an electrolytic cell supplied continuously with chemical materials stored outside the cell which react together to provide the chemical energy for conversion to electrical energy. In order for this to happen the materials must react within the cell, one material at the positive electrode and the other material at the negative electrode. Moreover, they must be prevented from reacting directly, for this would produce a kind of chemical short circuit, when no electrical energy would result from the reaction but usually only thermal energy would appear.

There are many types of chemical reaction suitable for this kind of cell but most attention has been focussed on combustion reactions. This is probably because there appears to be an almost unlimited supply of one of the necessary reactants—oxygen—and there is, of course, a wide variety of examples of the other component, many of which can be regarded as conventional fuels, that is to say, materials whose combustion will produce heat energy for conversion either directly or indirectly into mechanical work. It is for this reason that these devices are called fuel cells, although there is no *a priori* reason why they should not operate for reactions which are not those of the ordinary combustion of fossil fuels or materials derived from fossil fuels.

It would no doubt be possible in principle to construct a fuel cell whose chemical reaction was the combination of hydrogen and fluorine to form hydrogen fluoride, but the practical difficulties would be immense. Not only would gaseous hydrogen and gaseous fluorine react with each other with extreme violence, but also fluorine is likely to attack other materials in the cell (such as the electrodes) with great ease. Quite apart from that, a great deal of energy would need to be expended to produce the gaseous fluorine and the gaseous hydrogen from materials naturally containing them.

The most rewarding kind of reaction which could be used in a fuel cell would probably be that between a naturally occurring fossil fuel (such as a hydrocarbon oil or gas) and air. In other words, exactly the sort of reaction used to produce heat energy in the conventional engine or generating station.

Electrode processes

There are problems, as we shall see, associated with ensuring that the reaction does take place in the electrolytic cell: if it does not, electrical energy cannot be obtained directly. It is essential that an electron transfer reaction occurs at the surface of both electrodes, or else clearly no electric potential can arise between them, and the task of making certain that such processes actually happen is one of the big problems associated with fuel cells.

We are familar with certain kinds of so-called reversible electrode reactions. These are ones where, for example, a metal and its ions in solution are in equilibrium across the electrode surface, and the interaction between them may be regarded as proceeding through a series of stages each close to equilibrium. Cases where the mechanism is less obvious are those where a gas meets the electrode surface at a point where ions derived from the gas are present in the solution. A particular case of this general description might be the hydrogen electrode where the reaction

$$H_2(g) \rightarrow 2H^+(aq) + 2e^-$$

is set up for an aqueous electrolyte, or, say, the chlorine electrode, which involves the reaction

$$Cl_2(g) + 2e^- \rightarrow 2Cl^-(aq)$$

In each case the metal electrode (which might for example be platinum) is not only a current or potential collecting device but also clearly a surface active catalyst for the chemical process, and these processes can only take place at the boundary of the three phases.

The ease of setting up such systems depends therefore to a very large extent on the metal chosen for the electrode material. The best metal for the hydrogen electrode is probably platinum but the same metal would not necessarily be the most satisfactory for the oxygen electrode reaction

$$O_2(g) + 2H_2O(\ell) + 4e^- \rightarrow 4OH^-(aq)$$

The processes described so far refer only to the *reversible* electrode reactions occurring when no current is being taken from the cell and the potential difference across the electrodes is the *electromotive force* (emf). Nearly all applications of electrochemical cells do not fulfil this condition and currents (often substantial) are taken from them.

In those circumstances there is no equilibrium between electrode material or gas in contact with it and the ions in solution, but the reaction will definitely go only in one direction; in the cases we are considering ions will be formed which will then travel across the cell, carrying the current. The potential difference across the electrodes will be lowered because of the energy required to move the ions or because of the energy taken up in the various chemical rearrangements occurring at or near the electrode surfaces.

The various explanations of this phenomenon, sometimes called *polarisation*, will be discussed in chapter 4, but we can say here that in general the greater the current taken from the cell, the greater will be this polarisation or lowering of the working potential difference. Since power is the product of potential difference and current, this means that in some cases a high current will yield a lower power from a cell than would a smaller one. Moreover, a point may be reached when the current can be increased no further since it is kept to a limiting value by the polarisation effects—the energetics of the chemical reactions or the speed of movement of the ions.

Choice of cell reaction

From this brief account we can see that it would be quite easy to set up a cell whose electrode reactions were those of conversion of hydrogen and oxygen gases to hydrated protons and hydroxide ions respectively. A schematic diagram of this cell is shown in figure 1.1 Two cases are possible: one with an acidic electrolyte, containing more hydrated protons than hydroxide ions, and the other with an alkaline electrolyte of the opposite characteristics. The electrode reactions can be seen to be

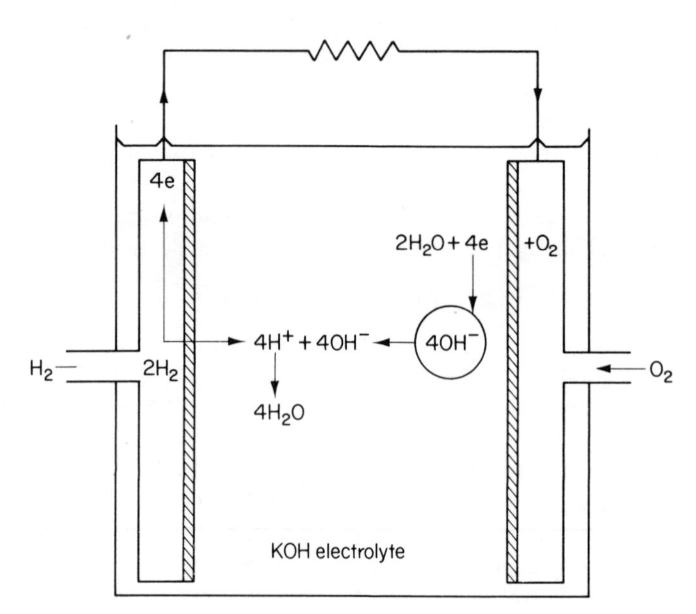

Figure 1.1 Reactions in hydrogen–oxygen fuel cell

for the acid electrolyte

$$H_2(g) + 2H_2O(\ell) \rightarrow 2H_3O^+(aq) + 2e^-$$

$$\tfrac{1}{2}O_2(g) + 2H_3O^+(aq) + 2e^- \rightarrow 3H_2O(\ell)$$

and *for the alkaline electrolyte*

$$H_2(g) + 2OH^-(aq) \rightarrow 2H_2O(\ell) + 2e^-$$

$$\tfrac{1}{2}O_2(g) + H_2O(\ell) + 2e^- \rightarrow 2OH^-(aq)$$

The overall effect is represented by what is undoubtedly the simplest combustion reaction

$$H_2(g) + \tfrac{1}{2}O_2(g) \rightarrow H_2O(\ell)$$

disregarding the small concentration of ions present in neutral water. Hydrogen is, of course, a fuel but there are others more abundant; these are more complicated chemically and it proves much more of a problem to find an electrode system capable of dealing with the direct conversion of, say, ethylene to some related ionic system in solution. Apart from the sheer chemical complexity of such reactions, suitable catalysts may well not exist. To illustrate the kind of chemical reaction we might be dealing with we can formulate the conversion of methane to bicarbonate ions and hydrated protons in aqueous solution at the electrode thus

$$12H_2O(\ell) + CH_4(g) \rightarrow HCO_3^-(aq) + 9H_3O^+(aq) + 8e^-$$

This is, of course, the overall reaction and should not be taken as any more than a guide to the process occurring at the fuel electrode. (Throughout this book the two electrodes are referred to as 'fuel' and 'oxidant' rather than by using the terms 'negative' and 'positive' (or 'cathode' and 'anode') since these are ambiguous, depending on whether the viewpoint is from the internal or external circuit).

However, each individual stage may well be more complicated than the relatively straightforward processes occurring at the hydrogen and oxygen electrodes. The provision of higher temperatures can ease this problem, since partial breakdown of hydrocarbon molecules on metal surfaces can well be facilitated by increasing the temperature. Another way of assisting the necessary intimacy of the various molecules is to use a hydrocarbon fuel or related substance which is soluble in the medium used (such as water). This accounts for the interest in fuel cells where the fuel, or material to be oxidised, is methanol or formaldehyde, both of which are readily soluble in water:

$$CH_3OH(aq) + 9H_2O(\ell) \rightarrow HCO_3^-(aq) + 7H_3O^+(aq) + 6e^-$$

$$HCHO(aq) + 7H_2O(\ell) \rightarrow HCO_3^-(aq) + 5H_3O^+(aq) + 4e^-$$

Thus we can see that a fair variety of reactions is available in principle, even if we only consider combustion processes. A superficial examination of the possibilities suggests that the simpler (in the chemical sense) the material chosen as fuel, the less intractable the problems encountered in setting up a satisfactory fuel cell system.

For example, it is clear from the preceding discussion that materials such as hydrogen, methanol and formaldehyde will make more satisfactory fuels for direct energy conversion than hydrocarbons, particularly those containing more than one carbon atom. Hence considerable interest is attached to methods of producing these relatively simple materials from naturally occurring hydrocarbons. Extensive details of such processes are out of place in a work of this sort but a summary of the possibilities is worth presenting. The production of hydrogen gas from methane, a process known as 'reforming', has been used since the nineteen thirties in the United States, and since the late nineteen fifties a similar process has been used to produce hydrogen from light naphtha in Britain. Generally, the hydrocarbon is mixed with steam and passed over a heated catalyst containing—amongst other materials—nickel (II) oxide, calcium oxide, silica and alumina. For methane, the chief reaction is

$$CH_4 + H_2O \rightarrow CO + 3H_2$$

and this equilibrium is shifted to the right at higher temperatures. The position is more complicated for higher hydrocarbons, particularly since it is considered important to prevent (as far as possible) any carbon being formed, since this may deactivate the catalyst.

Carbon monoxide produced by reforming may be readily converted to methanol with a zinc oxide/chromia catalyst

$$CO + 2H_2 \rightarrow CH_3OH$$

and this material can be dehydrogenated over silver or copper to produce formaldehyde

$$CH_3OH \rightarrow HCHO + H_2$$

All the fuel cell arrangements we have considered in this section have assumed the use of aqueous electrolytes, but this is certainly not essential. Provided that the substance separating the electrodes has electrolytic conductivity and can contain the ions produced at each electrode, then it is suitable. Ionic materials dissolved in non-aqueous solvents are rather rarely encountered (probably because water is by far the cheapest and most readily available ionising solvent), but molten salts of various types and even solid electrolytes have been tried out in various applications. The choice will depend, of course, to a large extent on the working temperature selected for the cell.

Fuel cell operation

We have seen that the oxidation of fuels, particularly chemically simple materials, can be carried out in principle more efficiently in a suitable galvanic cell than it can by a conventional combustion reaction followed by conversion of the thermal energy produced into electrical work. In itself this indicates a good reason for interest in fuel cells, but there are others worth considering. Some of these can best be dealt with in the chapter on applications (chapter 11) but it is useful to mention the main points here.

The potential difference produced by a single galvanic cell of any sort is usually quite small (1–2 V) and this means that production of high potentials requires a battery of such cells linked in series. High power installations may require cells to be connected in parallel. However since efficiencies of fuel cells depend in no way on the size of the battery of cells used, or indeed on the size of individual cells, neither of these types of arrangement is likely to be disadvantageous, which means that fuel cells are suitable for both low power and high power systems. This is of course in contra-distinction to the steam powered turbine generation of electricity which in general seems to increase in efficiency the larger the plant.

There are other economic advantages of fuel cells. It is unlikely, for example, that their manufacture would require any exceptional type of production process; the precision needed for construction of the various components is unlikely to be very great. Mechanical strength is not going to be very significant, since the cell itself will contain no moving parts and will not be subject to any great stresses. Constructional problems may arise for high temperature cells, as we shall see, but these will be of a different type and, again, will not require any great precision of method.

Maintenance of fuel cells also presents few problems, except perhaps in special cases where corrosion, for example, may present difficulties. Again, the lack of moving parts is significant here. It is instructive to realise that many working tests of fuel cell systems have only been ended by failure of peripheral equipment which is mechanical in nature, such as pumps for circulating the electrolyte or compressors for the gases used.

Finally, the requirement of invariance is important. This means that the fuel cell should act only as a converter of energy throughout its life and that the electrode materials and electrolyte should as far as possible be invariant. This is of course a clear distinction from the conventional battery and requires no side reactions or corrosion to occur.

It is often necessary to strike a balance between invariance and reactivity in the design of fuel cells, since characteristics affecting one favourably may well affect the other in the opposite sense.

Historical development of the fuel cell

The fuel cell is certainly not a recent invention. In 1802, Sir Humphrey Davy reported the construction of a simple cell whose reaction seemed to be

$$C + O_2 \rightarrow CO_2$$

and which would be written according to modern practice as

$$C(s)\,|\,H_2O(\ell),\ HNO_3(aq)\,|\,O_2(g)\,|\,C(s)$$

He was able to give himself a feeble electric shock from a battery of such devices. A more obvious fuel cell was that demonstrated in 1839 by Sir William Grove who succeeded in reversing a conventional electrolysis experiment—where water was broken down into hydrogen and oxygen by passing an electric current between two platinum electrodes immersed in sulphuric acid—and producing an electric current by supplying gaseous hydrogen and oxygen. In fact, by using a battery of twenty-six cells he was able to produce enough current to decompose water by electrolysis in yet another platinum–sulphuric acid system. It is interesting to note that Grove, from the very first, realised the essential problem of a working fuel cell—the difficulty of getting enough reaction to occur at the electrodes for high currents to be taken from the system. This is, of course, still a significant problem today.

The first use of the name 'fuel cell' seems to have been in 1889 by Mond and Langer who, using a similar contrivance to Grove, obtained current densities of about 0.2 A cm^{-2} which compares quite favourably with modern experience (current density means the current obtained per unit surface area of the electrodes). However, about this time was developed the dynamo, a device for converting mechanical work into electrical energy, and the success of this invention caused interest in fuel cells to lapse for almost 60 years. Another reason which has been adduced for this failure to consider fuel cells further was the development of the thermodynamic side of electrochemistry rather than the kinetic aspects of electrode reactions.

Modern interest in the fuel cell probably started with the work of F. T. Bacon in the nineteen fifties, culminating perhaps in the production of a 6 kW power plant of high pressure hydrogen–oxygen cells or in the subsequent development by an American company of his system to provide electrical power in the Apollo spacecraft. Details of different kinds of cells will be found in chapters 5–8, and chapter 10 contains details of operation. In discussing the use of fuel cells we must consider the thermodynamic basis of the energy conversion process (dealt with in chapters 2 and 3), but we must also make clear the dependence of successful working cells on the understanding of the kinetic processes involved—particularly of electrode reactions—and these polarisation effects will be elaborated in chapter 4.

CHAPTER 2

THE THERMODYNAMIC EFFICIENCY OF FUEL CELLS

The methods of chemical thermodynamics prove useful in the study of fuel cells despite the fact that normally only equilibrium properties can be discussed. There are two main avenues which can be explored. First, a study of the thermodynamics of the Carnot cycle for a perfectly reversible heat engine will show that there is a limiting efficiency which restricts the usefulness of any heat engine (as was mentioned in the previous chapter), and this can be compared with the possible efficiency of a fuel cell. Secondly, the electromotive force of a fuel cell (that is the maximum potential difference across the electrodes of a cell when no current is being taken from it) can be calculated by considering the electrochemical equilibrium set up when a fuel cell is studied potentiometrically.

The First Law of Thermodynamics

The First Law, or Principle, of Thermodynamics states that the total amount of all forms of energy—chemical, heat, mechanical, light and so on—contained by an isolated system after any physical or chemical change which does not involve a nuclear reaction or process, is exactly the same as it was before that change. In other words, it is simply the Conservation of Energy. An *isolated system* is one where neither mass nor energy can be lost to, or gained from, its surroundings. It may be subdivided into a *closed system* and its surroundings. A closed system is one within which there is conservation of mass, that is, mass cannot be transferred to or from it (but energy can). We have to exclude nuclear processes from consideration since they may, and frequently do, involve the interconversion of mass and energy according to the relation put forward by Einstein. Alternatively, we could formulate the law as the conservation of mass plus energy, which would always be obeyed but is perhaps less helpful in the present situation.

Having stated the First Law, we should now define certain quantities. The total energy of a closed chemical system is best represented by the *internal energy*, U, and changes in it either by ΔU (for a finite change) or

δU (for an infinitesimal change.) In both cases a positive change is taken as meaning one where the energy of the system concerned increases and, contrariwise, a negative change implies a decrease in the energy of the system. From the First Law, clearly changes in U are equal to the algebraic sum of the changes in all the other forms of energy. Thus for most chemical situations, we may write

$$U = q - w$$

where q represents the heat absorbed by the system, and w the work done by the system (which may commonly be mechanical or electrical). Sometimes we may need to consider an infinitesimal change

$$\delta U = \delta q - \delta w$$

where the symbols have similar meanings.

A rigorous application of the methods of chemical thermodynamics would show that U, the internal energy, is what is called a *function of state.* That is to say, its value depends only on the particular state of the system (its pressure and temperature and the amounts of various chemical species, for example) and not on the route by which the system reached that state. This has the important implication that the value of ΔU (or of the change in any other function of state) is independent of the manner in which the process takes place; that is, it is *path independent.* Such a characteristic is not true of q or w; they are *path dependent* and cannot be separately represented as a difference of initial and final state functions.

The most likely forms of work done by chemical reactions are the mechanical work performed by the expansion of gases and the electrical work done when the chemical reaction is arranged to form an electrolytic cell. The formulation of expressions for these quantities will depend on the route by which they occur. Here it is necessary to define thermodynamic reversibility, a concept of great theoretical utility. A process is said to be thermodynamically reversible if the system is, at all times during the process, only an infinitesimal distance away from equilibrium. For example, if physical process of gas compression is achieved by confining the gas in a cylinder closed by a piston subject to a pressure only very slightly greater than the pressure of the gas inside (see figure 2.1), then this process will be thermodynamically reversible. It follows that this example is an unreal one: gas compression could not be achieved in practice by this means even if a frictionless piston could be obtained. What then is the point of inventing such a process? The answer lies in the simplicity of the formula obtained for the work done (compared with the probably complicated formula necessary for an irreversible change), and that having once obtained the expression for ΔU for a reversible change we know for reasons mentioned above that this value must also apply for a corresponding irreversible (or 'real') change between the same two states.

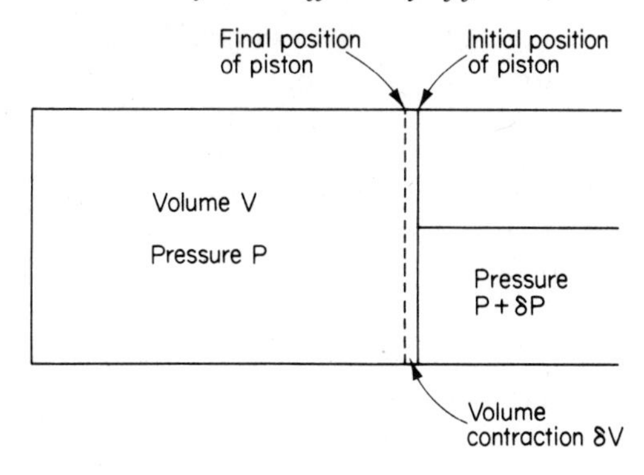

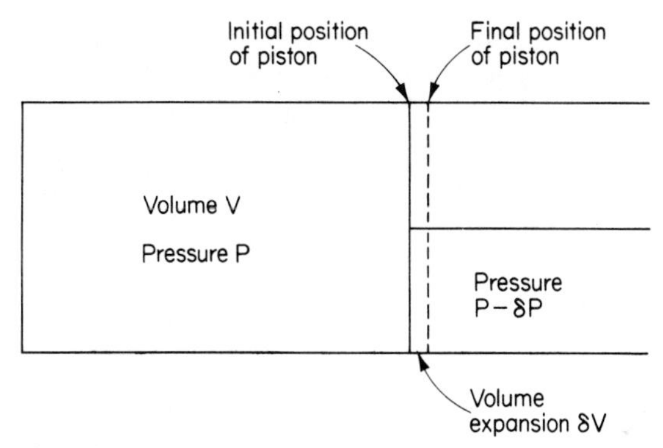

Figure 2.1 Thermodynamically reversible expansion of a gas

The formula for the work done in reversible gas expansion can be seen from figure 2.1: the expansion effected by a pressure outside the piston of $(P - \delta P)$ on a gas confined originally in a volume, V, will result in a change in volume of δV, to $(V + \delta V)$. Thus the work done by the gas will be $P\delta V$ (or the work done on the gas is $-P\delta V$). Summing this over a large number of such steps we have by the methods of integral calculus the expression

$$w_m = \int_{P_1,V_1}^{P_2,V_2} P\mathrm{d}V$$

where w_m is the mechanical work done by a gas being expanded from a state P_1, V_1 to a state P_2,V_2.

The case of a chemical change yielding electrical work is simpler; this is because the method generally used to measure the electromotive force of a galvanic cell allows the chemical change to be carried out by what is close to a thermodynamically reversible process. A potentiometric experiment essentially comprises a galvanic cell balanced against a standard cell (or a maintaining circuit calibrated by a standard cell). As the arrangement of resistances determining the balance point is shifted one way or the other, the chemical equilibrium in the cell under test is also shifted one way or the other—it either drives or is driven by the maintaining circuit. In the immediate vicinity of the balance point, therefore, the process can only be very slightly removed from equilibrium. The work done by the chemical cell under these conditions is hence simply given by the expression

$$\delta w_e = E\delta Q$$

where δw_e is the element of electrical work done by the cell whose emf is E when a quantity of charge δQ passes from high to low potential.

We can now write

$$\delta U = \delta q - P\delta V - E\delta Q$$

If volume is kept constant and there is no electrical work done, then $\delta U = \delta q_V$, or $\Delta U = q_V$ for a real change. In other words, ΔU represents the heat of reaction at constant volume. The heat of reaction at constant pressure, q_P, can be proved to be equal to ΔH, the change in enthalpy of the system. H is another function of state and is related to U by means of the equation

$$U = H - PV$$

which leads to the relation

$$\Delta U = \Delta H - P\Delta V$$

for a change at constant pressure.

It can be fairly simply shown that the work done by a gas on expansion in a thermodynamically reversible manner is the maximum work obtainable from the system: any process other than a perfectly reversible one will yield a smaller amount of work. Similar remarks apply to the production of electrical work. Thus

$$w_{max} = w_{rev}$$

This quantity may be rewritten as $-\Delta A$, where A is the Helmholtz function (or Helmholtz free energy):

$$\Delta A = -w_{rev}$$

A is defined in terms of U and the entropy, S

$$A = U - TS$$

A parallel state function is the Gibbs function (or Gibbs free energy), G, which is related to the 'net work' done by the system, that is the maximum work done neglecting any work of gas expansion:

$$w_{max} = w_{net} + \int P\mathrm{d}V$$

$$\Delta G = -w_{net}$$

The Gibbs function is defined in terms of the entropy and the enthalpy of the system:

$$G = H - TS$$

A particularly important result is that for an electrochemical cell working reversibly at constant pressure and temperature; the electrical work done is the net work and hence $\Delta G = -w_e$. For this to be a reversible process, it is necessary for the cell to be on 'open circuit', that is, no current flows and the potential difference across the electrodes is the emf, E. Thus $\Delta G = -zFE$ where F is the Faraday constant (the product of the Avogadro constant and the charge on the electron) and z is the number of electrons transferred when the cell reaction proceeds from left to right.

The Second Law of Thermodynamics

The idea of entropy—a quantity often used, though difficult to explain satisfactorily—arises from the Second Law of Thermodynamics. Whereas the First Law effectively states that all forms of energy are equivalent and can be converted from one to another, the Second Law imposes restrictions on this interconversion. There are many forms of statement of the Second Law and, unhappily, few seem at first sight relevant to the question of chemical reactivity which is the fundamental point about the application of chemical thermodynamics. Two of the better known statements are: 'It is not possible to transfer heat spontaneously from a cold body to a warmer one without doing work on the system', and 'For a spontaneous change in an isolated system the entropy of the system always increases'.

The first is helpful only in the working of heat engines, and the second introduces problems of explaining entropy and of confining our conclusions to isolated systems (ones where transfer of material or energy to or from the surroundings is prohibited), which are not commonly encountered in chemistry.

The 'free energy' functions A and G were invented specifically to provide more chemically useful statements of the Second Law; they are complementary and they provide criteria of equilibrium for systems at constant temperature. For a closed system at constant volume and temperature undergoing a spontaneous change, A must decrease, and for such a system at constant pressure and temperature undergoing a spontaneous change, G must decrease. Thus, at equilibrium the values of A and G must remain constant, or $(\delta G)_{P,T} = 0$ and $(\delta A)_{V,T} = 0$. These criteria of equilibria can be shown to be consistent with the Second Law by considerations of the definitions of A and G.

The definition of entropy arises from the heat engine type of system. It can be shown by a logical scheme known as Carnot's theorem that the quantity (q_{rev}/T), where q_{rev} is the heat change for a reversible process, is a constant for *all* changes from the initial to the final state, that is, the quantity is a function of state. In fact it is the increase in entropy for the change

$$\Delta S = S_B - S_A = \sum_A^B (q_{rev}/T)$$

or, for an infinitesimal change

$$\delta S = \delta q_{rev}/T$$

For any spontaneous irreversible change

$$\delta S > \delta q_{irrev}/T$$

and this allows us to see the basis for our second statement of the Second Law.

This definition of entropy does not, of course, explain it. It is necessary to go into the field of statistical or molecular thermodynamics to do that, and we find that the entropy of a system can be related to the number of 'microstates' or 'complexions', W, accessible to molecules in the assembly by the equation

$$S = k \ln W$$

where k is the Boltzmann constant (sometimes called the gas constant per molecule). W may also be thought of as the number of ways the molecules of the system can be distributed among the various energy states, and hence entropy is often said to depend on the disorder or 'randomness' of the system. Unfortunately this last statement is not only imprecise as it stands but it is also often interpreted as being confined to spatial disorder, that is, the arrangements of atoms within a molecule, or of molecules within an assembly.

The Third Law of Thermodynamics

To a chemist the evaluation of entropy changes is usually only important in the computation of changes in Gibbs function so that positions of equilibrium may be determined. In such an evaluation the Third Law of Thermodynamics is usually employed; one form of this law is that all perfect crystalline substances at the absolute zero have zero entropy. We do not need to consider this further.

The Carnot cycle

The efficiency of a heat engine—that is, a device for converting the thermal energy of a system at a high temperature into mechanical work—is best assessed in terms of the *Carnot cycle.* This cycle is a sequence of thermodynamically reversible operations carried out on a gas between two temperatures, the final state being identical with the initial.

An amount n^1 of an ideal gas is caused to undergo a series of expansions

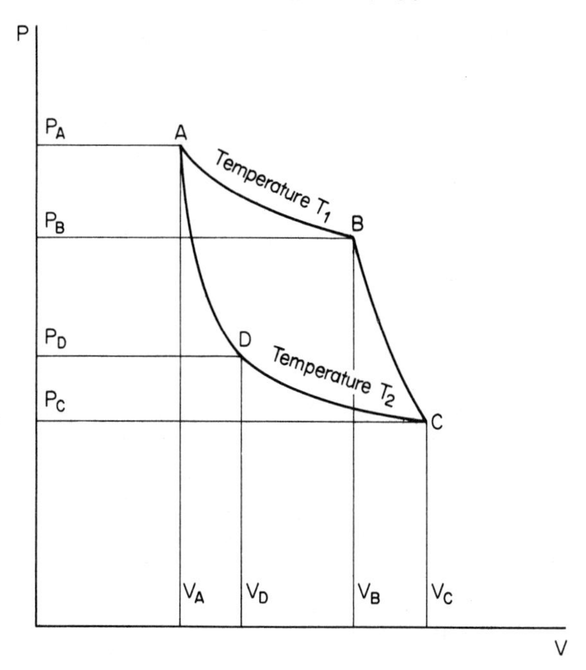

Figure 2.2 Pressure–volume relations for the Carnot cycle

and compressions, ABCD, as indicated in figure 2.2. The gas is initially at temperature T_1, and has a volume V_A. It expands isothermally and reversibly to a volume V_B, and then reversibly and adiabatically from V_B to V_C, its temperature falling from T_1 to T_2. An adiabatic change is one where no heat is gained from or lost to the surroundings.

The gas is compressed from V_C to V_D reversibly and isothermally at the temperature T_2, and finally the gas is brought back to its starting point by a reversible adiabatic compression from V_D to V_A.

The heat absorbed at the higher temperature is

$$q_{AB} = nRT_1 \ln \frac{V_B}{V_A}$$

and that lost at the lower is

$$q_{CD} = nRT_2 \ln \frac{V_C}{V_D} = nRT_2 \ln \frac{V_B}{V_A}$$

It is clear that $w = q_{AB} - q_{CD}$ and hence $\Delta U = 0$ which of course fulfils the Conservation of Energy principle of the First Law.

Efficiency

The efficiency (η) of the heat engine working in such a cycle, is the ratio of the work done during the cycle to the heat 'put in'— that is, absorbed at the higher temperature—and this can quite readily be shown to be given by the expression

$$\eta = \frac{nRT_1 \ln (V_B/V_A) - nRT_2 \ln (V_B/V_A)}{nRT_1 \ln (V_B/V_A)}$$

$$= \frac{T_1 - T_2}{T_1}$$

This very simple expression represents the ideal efficiency (and hence its maximum value) of *any* heat engine working between the temperatures T_1 and T_2. The proof of this assertion is to be found in Carnot's theorem. It explains the importance of using as high a temperature as possible for the steam employed in driving turbines in a conventional power station, and also the use of steam condensers where possible (usually using river water) to keep the value of T_2 low. Some representative values of this optimum efficiency may be of interest. If, for the purposes of simple calculation, we take the lower temperature as 300 K (27 °C, somewhat above ambient temperatures in most parts of the world), then an engine working with steam at 400 K (127 °C, a little above the boiling point of water at ordinary pressures) can have an efficiency no greater than $(400–300)/400 = 0.25 = 25$ per cent. That is to say 75 per cent of the heat energy of the steam used is not available for conversion to useful work. Increasing the steam temperature to 600 K produces an increase in maximum efficiency to 50 per cent, and the use of steam at 3000 K would bring it up to 90 per cent. This suggests that use of steam superheated to moderately high temperatures (say 600–1000 K or about 300–700 °C) would produce a worthwhile increase in the possible efficiency.

It is important to stress again that these figures represent only the best possible efficiencies of heat engines and they can never be realised in practice. The actual working efficiencies must always be lower, in some cases much lower, because of such factors as heat wastage and frictional losses. Since such losses may become greater at higher working temperatures the theoretical increase in efficiency referred to earlier may be illusory, but in fact there is always some advantage in practice to be gained from operation at the higher temperatures, as can be seen from the observation that most modern conventional electricity power stations are built to use steam at about 550 °C.

Efficiency of a fuel cell

How can we assess the optimum efficiency of a fuel cell system in order to compare with a heat engine? In chapter 1 we made it clear that fuel cells were devices working at constant temperature as far as this was possible (that is isothermally) and hence they are not subject to the restrictions of energy conversion we have just outlined for Carnot systems. The efficiency must be defined in a similar way, as the ratio of energy put in to useful work taken out, but how are we to define the energy put in?

If we take the energy supplied as the chemical or 'free' energy of the fuel cell reaction, then it ought to be the decrease in Gibbs function for the chemical system; since this is equal to the electrical work obtained when no current flows (as we have shown earlier) then the optimum efficiency on this basis of all fuel cells is 100 per cent. This is not perhaps the most helpful comparison, and in order to make a more useful one we might take the enthalpy decrease ($-\Delta H$) for the chemical reaction as representing the energy put in. Such a scheme, while slightly artificial, would certainly afford direct comparability, since this enthalpy change is the heat energy supplied when fuels are burnt in the conventional manner to raise steam and generate electricity by mechanical means. Once again, we take the decrease in Gibbs function ($-\Delta G$) as being equal to the maximum electrical energy obtained. Thus the efficiency (η) is given by

$$\frac{-\Delta G}{-\Delta H}$$

for any particular fuel cell system.

Earlier we showed that $\Delta G = \Delta H - T\Delta S$ (at constant temperature): we may write

$$\eta = \frac{\Delta H - T\Delta S}{\Delta H} = 1 - \frac{T\Delta S}{\Delta H}$$

In order to make proper comparison of one chemical system with another we should use the changes in the *standard* molar thermodynamic quantities, $\Delta G^{\ominus}$, $\Delta H^{\ominus}$ and $\Delta S^{\ominus}$. This is because the value of, for example, the change in Gibbs function for a particular reaction depends on the amounts and concentrations of the various reactants and products concerned, and it is sensible to standardise these to provide a common basis for comparison. This idea will be elaborated in the next chapter.

Thus

$$\eta = \frac{\Delta G^{\ominus}}{\Delta H^{\ominus}} = 1 - \frac{T\Delta S^{\ominus}}{\Delta H^{\ominus}}$$

Values of the changes in standard entropy and enthalpy can be obtained from compendia of thermodynamic data for various chemical changes at various temperatures, and the efficiencies calculated. These efficiencies are again hypothetical maximum efficiencies since, again, operating losses (and any deviation from standard state conditions) are likely to reduce their value. Some representative figures for simple reactions are shown in table 2.1.

The artificial basis of these calculated efficiencies is responsible for the values of greater than 100 per cent obtained for reactions with negative

Table 2.1 Efficiencies of some simple reactions

	T/K	$\Delta H^{\ominus}$/kJ mol^{-1}	$\Delta G^{\ominus}$/kJ mol^{-1}	efficiency (per cent)
$H_2(g) + \frac{1}{2}O_2(g) \rightarrow H_2O(g)$	298	−241.7	−228.5	94.5
	500	−243.7	−219.0	89.9
	1000	−247.6	−192.5	77.7
	2000	−252.0	−135.1	53.6
$H_2(g) + \frac{1}{2}O_2(g) \rightarrow H_2O(\ell)$	298	−285.8	−237.2	83.0
$CH_4(g) + 2O_2(g) \rightarrow CO_2(g) + 2H_2O(g)$	298	−889.9	−817.6	91.9
$CH_3OH(g) + \frac{3}{2}O_2(g) \rightarrow CO_2(g) + 2H_2O(g)$	298	−718.9	−698.2	97.1
$N_2H_4(g) + O_2(g) \rightarrow N_2(g) + 2H_2O(g)$	298	−605.6	−601.8	99.4
$C(s) + \frac{1}{2}O_2(g) \rightarrow CO(g)$	298	−110.53	−137.16	124.1
$CO(g) + \frac{1}{2}O_2(g) \rightarrow CO_2(g)$	298	−282.8	−257.0	90.9
$C(s) + CO_2(g) \rightarrow 2CO(g)$	298	+172.1	+119.6	69.5

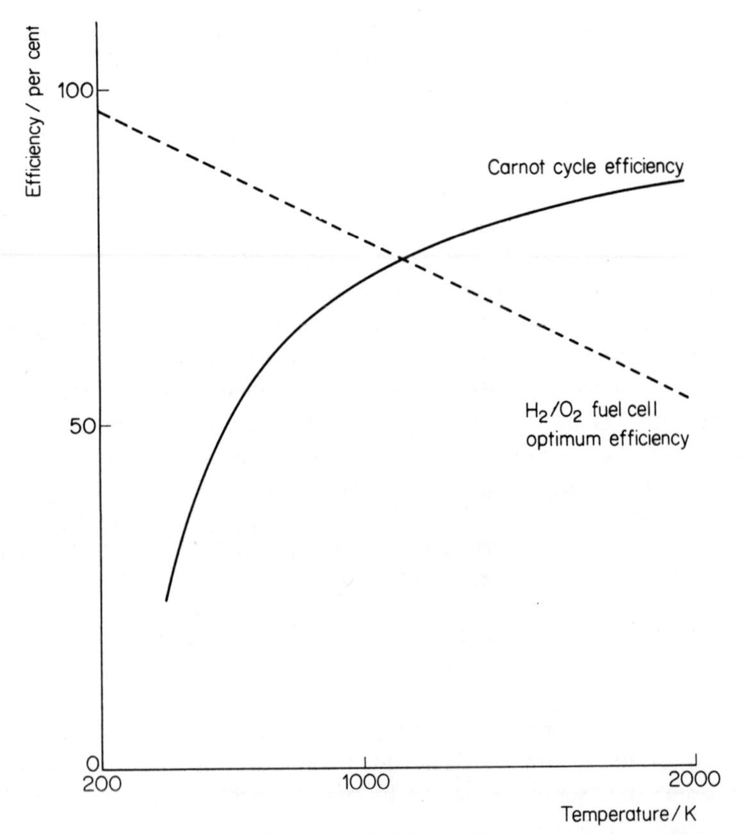

Figure 2.3 Comparison of Carnot cycle efficiency (lower temperature = 30 K, upper temperature indicated on x-axis) and ideal efficiency of hydrogen–oxygen fuel cell

standard entropy changes and positive standard enthalpy changes. An example is the oxidation of carbon to carbon monoxide. These values arise because the system being examined is no longer a closed one and energy is obtained from the surroundings; this energy is normally thermal and is numerically equal to $T\Delta S^{\ominus}$ for the reactions quoted. We have assumed this basis largely in order to provide comparison with heat engine efficiencies. A graphical comparison is shown in figure 2.3 where the optimum efficiency of the hydrogen–oxygen fuel cell system working at various temperatures is contrasted with the optimum efficiency of a heat engine working between a fixed lower temperature and various higher temperatures.

It can be seen from figure 2.3 that the hydrogen–oxygen reaction would provide a fuel cell system potentially far superior in efficiency to any conventional heat engine system at temperatures below about 850 °C. Above that temperature the fuel cell system shows no advantage; in fact the conventional power station would probably be more efficient. We must realise that these are ideal efficiencies—the actual working efficiencies may not follow the same pattern—and that the fuel cell efficiency is very much dependent on the chemical reaction being used, as, of course has been shown by the figures in table 2.1.

Fuel cell efficiencies are reduced below the values we have quoted by the actual operation of the cell when current is taken from it, and the reasons for these changes are discussed in chapter 4. The rates of electrode processes and their influence on the current and potential of the cell are also dealt with in this chapter.

There are several ways of expressing the working efficiency of fuel cells (or of any other electrochemical cells, for that matter). One way is to take the ratio of work obtained under operating conditions to the maximum work obtainable (which would of course be under 'open circuit' conditions). Thus

$$\eta_{\text{operating}} = \frac{E'dQ}{-\Delta G}$$

where E' is the operating potential difference, dQ is the charge transformed and ΔG is the change in Gibbs function for the cell reaction. ΔG has been shown to be equal to $-zFE$ and hence η becomes $E'dQ/zFE$. Such an expression is often referred to as the *energy efficiency* of the cell.

Notes

1. n is frequently, but imprecisely, referred to as the *number of moles* of the ideal gas. Further information may be found in *Physiochemical Quantities and Units*, by M. L. McGlashan, Chemical Society Monographs for Teachers, no. **15**.

CHAPTER 3

ELECTROMOTIVE FORCE OF FUEL CELLS

In the previous chapter we have shown that the change in Gibbs function for a fuel cell reaction—or the reaction associated with any other galvanic cell—is equal to the electrical work done (though of opposite sign), which in turn depends on the electromotive force, or emf, of the cell when it is working reversibly under conditions where no current is taken from it. Although such conditions seem irrelevant to the actual working state of the cell (when large currents may well be taken from it), it is nevertheless worth considering the factors influencing the emf since this will determine the optimum efficiency, as we have seen earlier. It is usually thought most convenient to consider working potential differences in terms of their deviations from the emf, and this provides a further reason for looking further at the open circuit situation. These deviations arise from the effect of taking a current from the cell and are usually considered as kinetic effects. A discussion of them will be found in chapter 4.

The emf of a galvanic cell is conveniently divided into two constituent electrode potentials. This process, while convenient, is entirely arbitrary: strictly the emf measured is the potential across two terminals of the potentiometer, and this is the sum of several potentials—one for each phase change in the system. For example, the Daniell cell may be represented as

$$Cu(s)\,|\,CuSO_4(aq)\,||\,ZnSO_4(aq)\,|\,Zn(s)$$

where metallic copper and zinc electrodes dip into aqueous solutions of copper (II) sulphate and zinc sulphate respectively. These solutions are prevented from mixing by a semipermeable membrane, in practice a porous porcelain container, which restricts the passage of certain ions. The measured emf of this cell is the algebraic sum of the following interphase potentials:

(1) Between Cu(s) and Zn(s).
(2) Between Cu(s) and $CuSO_4$(aq).
(3) Between $CuSO_4$(aq) and $ZnSO_4$(aq).
(4) Between $ZnSO_4$(aq) and Zn(s).

However, none of these can be measured independently and the conventional division into two electrode potentials (corresponding more or less to items 2 and 4) has some advantage. It is important to realise that there is no possible way of measuring an individual electrode potential, and therefore any division depends on the arbitrary selection of a zero value for one particular electrode system.

Standard electrode potentials

It is usually found most satisfactory to define a standard potential in terms of a standard concentration of the ion involved, since the emf of a cell depends on the concentrations of ions concerned in what is called the cell reaction. For example the Daniell cell reaction may be written as

$$Zn(s) + CuSO_4(aq) \rightarrow ZnSO_4(aq) + Cu(s)$$

and its emf depends on the concentrations of copper (II) and zinc ions in a way which will be shown later. If standard values of these concentrations are chosen, then the measured emf will be the *standard* emf and this also corresponds to the change in the *standard* thermodynamic functions ($\Delta H^{\ominus}$, $\Delta G^{\ominus}$, $\Delta S^{\ominus}$) for the cell reaction we have quoted.

The standard emf is considered as the difference of the two standard electrode potentials, and these are arranged on a scale where the standard hydrogen electrode potential is taken as zero at all temperatures. The hydrogen electrode is one where hydrogen gas is supplied to a platinised platinum plate dipping into an aqueous solution containing hydrogen ions; standard conditions fix the pressure of the gas and the concentration of the ions.

The sign of a standard electrode potential depends on the spontaneous direction of the electrode reaction with hydrogen when all concentrations and pressures are standard. This is best considered by constructing a hypothetical cell of the electrode concerned and the hydrogen electrode. For example

$$H_2(g) + Zn^{2+}(aq) \rightarrow 2H^+(aq) + Zn(s)$$

is the cell reaction for the combination of the zinc electrode with the hydrogen electrode; the spontaneous direction of this reaction is from right to left (metallic zinc 'dissolves' in aqueous acids), and we define its standard electrode potential as negative relative to the standard hydrogen electrode.

Similarly the standard potential of the chlorine electrode will be positive, since the reaction

$$H_2(g) + Cl_2(g) \rightarrow 2H^+(aq) + 2Cl^-(aq)$$

is spontaneous from left to right. This is the convention adopted since 1953 by the International Union of Pure and Applied Chemistry. It leads to the standard electrode potentials of the alkali metals being negative. An exactly opposite convention has in the past been adopted, in particular by American scientists, and this of course leads to positive alkali metal standard electrode potentials. Some values of standard electrode potentials defined in this way are given in table 3.1

Table 3.1 Standard electrode potentials in aqueous solution

Defining reaction*	$V^{\ominus}$/V
$Li^+(aq) + e^- \rightarrow Li(s)$	−3.045
$K^+(aq) + e^- \rightarrow K(s)$	−2.925
$Ba^{2+}(aq) + 2e^- \rightarrow Ba(s)$	−2.90
$Ca^{2+}(aq) + 2e^- \rightarrow Ca(s)$	−2.87
$Na^+(aq) + e^- \rightarrow Na(s)$	−2.714
$La^{3+}(aq) + 3e^- \rightarrow La(s)$	−2.52
$Al^{3+}(aq) + 3e^- \rightarrow Al(s)$	−1.66
$ZnO_2^{2-}(aq) + 2H_2O(\ell) + 2e^- \rightarrow Zn(s) + 4OH^-(aq)$	−1.216
$2H_2O(\ell) + 2e^- \rightarrow H_2(g) + 2OH^-(aq)$	−0.828
$Zn^{2+}(aq) + 2e^- \rightarrow Zn(s)$	−0.763
$S(s) + 2e^- \rightarrow S^{2-}(aq)$	−0.48
$Fe^{2+}(aq) + 2e^- \rightarrow Fe(s)$	−0.440
$PbSO_4(s) + 2e^- \rightarrow Pb(s) + SO_4^{2-}(aq)$	−0.356
$Ni^{2+}(aq) + 2e^- \rightarrow Ni(s)$	−0.250
$CO_2(g) + 2H^+(aq) + 2e^- \rightarrow HCOOH(aq)$	−0.196
$MnO_2(s) + 2H_2O(\ell) + 2e^- \rightarrow Mn(OH)_2(s) + 2OH^-(aq)$	−0.05
$2H^+(aq) + 2e^- \rightarrow H_2(g)$	0.000
$HCOOH(aq) + 2H^+(aq) + 2e^- \rightarrow HCHO(aq) + H_2O(\ell)$	0.056
$C(s) + 4H^+(aq) + 4e^- \rightarrow CH_4(g)$	0.13
$Cu^{2+}(aq) + 2e^- \rightarrow Cu(s)$	0.153
$HCHO(aq) + 2H^+(aq) + 2e^- \rightarrow CH_3OH(aq)$	0.19
$O_2(g) + 2H_2O(\ell) + 4e^- \rightarrow 4OH^-(aq)$	0.401
$I_2(s) + 2e^- \rightarrow 2I^-(aq)$	0.536
$CH_3OH(aq) + 2H^+(aq) + 2e^- \rightarrow CH_4(g) + H_2O(\ell)$	0.586
$Hg_2^{2+}(aq) + 2e^- \rightarrow 2Hg(\ell)$	0.789
$Ag^+(aq) + e^- \rightarrow Ag(s)$	0.799
$Br_2(\ell) + 2e^- \rightarrow 2Br^-(aq)$	1.065
$O_2(g) + 4H^+(aq) + 4e^- \rightarrow 2H_2O(\ell)$	1.229
$Cl_2(g) + 2e^- \rightarrow 2Cl^-(aq)$	1.360
$PbO_2(s) + SO_4^{2-}(aq) + 4H^+(aq) + 2e^- \rightarrow PbSO_4(s) + 2H_2O(\ell)$	1.685
$F_2(g) + 2e^- \rightarrow 2F^-(aq)$	2.65

*Strictly, $V^{\ominus}$ is the standard emf of a cell whose reaction involves the conversion of $H_2(g)$ into $H^+(aq)$ as well as the defining reaction quoted here. For example, the cell reaction for the standard electrode potential of lithium would be

$$Li^+(aq) + \tfrac{1}{2}H_2(g) \rightarrow H^+(aq) + Li(s)$$

Table 3.2 Emf values for systems of interest

Reaction	$\Delta G^{\ominus}$/kJ mol^{-1}	$E^{\ominus}$/V
$H_2(g) + \tfrac{1}{2}O_2(g) \rightarrow H_2O(g)$	−228.5	1.184
$H_2(g) + \tfrac{1}{2}O_2(g) \rightarrow H_2O(\ell)$	−237.2	1.229
$CH_4(g) + 2O_2(g) \rightarrow CO_2(g) + 2H_2O(g)$	−817.6	1.060
$CH_3OH(aq) + \tfrac{3}{2}O_2(g) \rightarrow CO_2(g) + 2H_2O(g)$	−698.2	1.206
$N_2H_4(g) + O_2(g) \rightarrow N_2(g) + 2H_2O(g)$	−601.8	1.559
$CH_3CHO(aq) + \tfrac{5}{2}O_2(g) \rightarrow 2CO_2(g) + 2H_2O(g)$	−1122.0	1.163
$Pb(s) + PbO_2(s) + 2H_2SO_4(aq) \rightarrow 2PbSO_4(s) + 2H_2O(\ell)$	−393.9	2.041

The standard emf of any cell system can be calculated by subtraction of the standard potentials for its two constituent electrode systems. Thus the standard emf of the system in which hydrogen and oxygen react to give water will be 1.229 V; values of emf for some other systems of interest are to be found in table 3.2. Both standard emf and standard electrode potential are referred to particular standard concentrations and pressures, as stated in the tables. The choice of these standards is discussed in the following section.

Dependence of emf on electrolyte concentration

We are now interested in evaluating the emf of cells which do not contain the component materials in their standard states (that is, at standard pressures or concentrations). In other words we want to know the way in which the emf of a cell varies with the concentrations of the various constituents, both ionic materials in the cell electrolyte and gaseous materials supplied to the electrodes.

The form of the variations can of course be established by experiment for certain systems, but it can also be derived theoretically by consideration of the molar change in Gibbs function for a cell reaction and its dependence on the partial pressures or concentrations of the reactants and products. Since we have already established the connection between Gibbs function change and electrical work done, the emf dependence can readily be obtained.

Suppose we take a general chemical reaction

$$aA + bB \rightleftharpoons yY + zZ$$

(a, b, y and z being the stoichiometric coefficients, that is the number of molecules, of A, B, Y and Z in the chemical equation). For each component of the reaction we can define a quantity μ, called the *chemical potential* such that

$$\Delta G = \mu_Y n_Y + \mu_Z n_Z - \mu_A n_A - \mu_B n_B$$

and hence for a small increment in G at constant temperature and pressure

$$\begin{aligned}\delta G &= \mu_Y \delta n_Y + \mu_Z \delta n_Z - \mu_A \delta n_A - \mu_B \delta n_B \\ &= \sum \mu_i \delta n_i\end{aligned}$$

where δn_A, for example, is the increase in amount of A. (It should be noted that chemical potential is defined strictly in terms of a partial differential

$$\mu_i = \left(\frac{\partial G}{\partial n_i}\right)_{P,T,n_j,n_k,\ldots}$$

but may be regarded as the contribution made to the total Gibbs function per unit amount of each substance present.)

At constant pressure and temperature, the molar change in Gibbs function for the reaction, ΔG, can be shown to be given by

$$\Delta G = y\mu_Y + z\mu_Z - a\mu_A - b\mu_B \tag{3.1}$$

(Note that this is an algebraic sum, with 'products' taken as positive and 'reactants' as negative. This fulfils the usual chemical convention that the materials on the left hand side of an equation are called reactants and those on the right hand side are called products, irrespective of the actual spontaneous direction of reaction.)

Standard methods of chemical thermodynamics can be used to define the *activity* a_i of any substance in terms of its chemical potential:

$$\mu_i = \mu_i^\ominus + RT \ln a_i$$

where $\mu_i^\ominus$ is the *standard chemical potential* of the species i, whose value is dependent on the choice of standard state conditions, which must of course also correspond mathematically to $a_i = 1$. Two cases concern us particularly: the gaseous reactant activity is related to *partial pressure*, P_i, and standard pressure, $p^\ominus$, by the equation

$$a_i = \frac{P_i}{p^\ominus} \gamma_i$$

and for solid solutes it is related to *molality* (amount of solute per unit mass of solvent) by

$$a_i = \frac{m_i}{m^\ominus} \gamma_i$$

where $m^\ominus$ is a *standard molality;* γ_i in both cases is called the *activity coefficient*. For an ideal system, which means one obeying the rules (the ideal gas laws for a gas, Henry's law for a solution), the activity coefficient will be unity.

The standard state in each case corresponds mathematically to $a_i = 1$ and this is commonly taken physically to mean for gases an ideal gas having a partial pressure of 1 atm, and for solutions an ideal solution of 1 mol of solute dissolved in 1 kg of solvent. In the case of the gaseous substance the standard state has physical reality but in the case of the solution it almost certainly does not, since solutions of such concentration are usually far from ideal in behaviour. However, since both states are—as their name implies—only used as *standards* to provide a scale of chemical potential and activity, whether their actual behaviour is ideal or not is unimportant.

The behaviour of electrolyte solutions is sometimes related to a standard state based on the *molarity scale*, which may be more convenient for preparation.

Thus here

$$a_i = \frac{c_i}{c^\ominus} \gamma_i$$

where $c^{\ominus}$ is the standard molarity (1 mol dm^{-3} is commonly taken). For aqueous solutions at ordinary temperatures the difference between 1 mol dm^{-3} and 1 mol kg^{-1} is slight, because 1 dm^3 of solution contains about 1 kg of solvent.

Unless otherwise stated, data tabulated in this book are based on standard state referred to the Molarity scale.

Equation 3.1 now becomes

$$\Delta G = y\mu_{\mathrm{Y}}^{\ominus} + z\mu_{\mathrm{Z}}^{\ominus} - a\mu_{\mathrm{A}}^{\ominus} - b\mu_{\mathrm{B}}^{\ominus} + RT \ln \frac{a_{\mathrm{Y}}^{y} a_{\mathrm{Z}}^{z}}{a_{\mathrm{A}}^{a} a_{\mathrm{B}}^{b}}$$

The algebraic sum of the various standard chemical potential terms is normally called the standard molar change in Gibbs function, $\Delta G^{\ominus}$, for the reaction. Thus

$$\Delta G = \Delta G^{\ominus} + RT \ln \left(\frac{a_{\mathrm{Y}}^{y} a_{\mathrm{Z}}^{z}}{a_{\mathrm{A}}^{a} a_{\mathrm{B}}^{b}} \right)$$

For the reaction which is involved in a galvanic or electrolytic cell this quantity is related to the electrical work done. This is perhaps best illustrated by considering a real case. The simplest fuel cell reaction is that of the hydrogen–oxygen reaction which we will write as

$$2\mathrm{H_2(g)} + \mathrm{O_2(g)} + 6\mathrm{H_2O}(\ell) \rightarrow 4\mathrm{H_3O^+(aq)} + 4\mathrm{OH^-(aq)}$$

The electrical work done by the cell is given by

$$w_e = zFE$$

as was established in chapter 2. This relation is only correct when the charge passes under conditions of thermodynamic reversibility, but since this is very close to the way in which emf is measured we can probably accept it. In the reaction we have specified, z will be equal to 4, since effectively four electrons are transferred from $2H_2$ to O_2. A consideration of the separate electrode reactions should support this assertion:

$$2\mathrm{H_2} + 4\mathrm{H_2O} \rightarrow 4\mathrm{H_3O^+} + 4\mathrm{e^-}$$
$$4\mathrm{e^-} + \mathrm{O_2} + 2\mathrm{H_2O} \rightarrow 4\mathrm{OH^-}$$

Thus $w_e = 4EF$ here. Since $\Delta G = -w_e$

$$\Delta G = -4EF$$

For this reaction ΔG is given by $\Delta G^{\ominus} + RT \ln (a_{\mathrm{H_3O^+}}^4 a_{\mathrm{OH^-}}^4 / a_{\mathrm{H_2}}^2 a_{\mathrm{O_2}} a_{\mathrm{H_2O}}^6)$ and since we may write $\Delta G^{\ominus} = -nE^{\ominus}F$ by analogy, where $E^{\ominus}$ is the standard emf of the cell (assuming of course that we take the same standard states for both $\Delta G^{\ominus}$ and $E^{\ominus}$) then the emf is given by

$$E = E^{\ominus} - \frac{RT}{4F} \ln \frac{a_{\mathrm{H_3O^+}}^4 a_{\mathrm{OH^-}}^4}{a_{\mathrm{H_2}}^2 a_{\mathrm{O_2}} a_{\mathrm{H_2O}}^6} \tag{3.2}$$

Such a relation between the emf of a cell and the activities of the substances in it is sometimes called the Nernst equation.

The Nernst equation

Activity, as we have seen, describes non-ideal behaviour of species with reference to a standard state based on a particular concentration scale. If we put in ionic activity coefficients to make this relation clearer, equation 3.2 would become

$$E = E^{\ominus} - \frac{RT}{4F} \ln \frac{m^4_{H_3O^+} \gamma^4_{H_3O^+} m^4_{OH^-} \gamma^4_{H_3O^-}/\text{mol}^8\,\text{kg}^{-8}}{P^2_{H_2} P_{O_2} a^6_{H_2O}/\text{atm}^3} \qquad (3.3)$$

or perhaps in a less cumbrous form

$$E = E^{\ominus} - \frac{2RT}{8F} \ln (m/\text{mol kg}^{-1}) + \frac{RT}{2F} \ln (P_{H_2}/\text{atm}) + \frac{RT}{4F} \ln (P_{O_2}/\text{atm})$$

where, since the ions are all in the same solution which is electrically neutral, we can put $m_{H_3O^+} m_{OH^-} = m$, and if the solution is fairly dilute we can put $a_{H_2O} = 1$ and the various activity coefficients equal to unity.

For gases at ordinary pressures this assumption does not sacrifice much accuracy, but for most electrolyte solutions of finite concentration it is unreasonable. It is generally held, moreover, that activities or activity coefficients of single ions have no physical reality in a rigorous thermodynamic description of this sort of system, and so *mean ionic activities* and *mean ionic activity coefficients* are used instead. For a 1:1 electrolyte, such as $NaCl \rightleftharpoons Na^+ + Cl^-$, the relations are

$$a^2_{\pm} = a_+ a_-$$

and

$$\gamma^2_{\pm} = \gamma_+ \gamma_-$$

where the quantities with subscripts + and − refer to single ions, and $a_{\pm}$ and $\gamma_{\pm}$ are the mean ionic activity and activity coefficient. (The definition is more complicated for anything other than 1:1 electrolytes.) These quantities represent the behaviour of both positive and negative ions taken together.

Equation 3.3 now becomes

$$E = E^{\ominus} - \frac{RT}{8F} \ln \frac{(m/\text{mol kg}^{-1})^2 \gamma^2_{\pm}}{P^2_{H_2} P_{O_2}/\text{atm}^3}$$

if we assume ideal gaseous behaviour, and unit activity of water.

In this way it is possible to calculate the dependence of emf on molality of the ions in the solution provided we know the value of $\gamma_{\pm}$ at the appropriate concentration. There are various methods, both by experiment and by calculation, for obtaining such values, and the reader is referred to textbooks on the behaviour of electrolyte solutions for details. In fact, in the particular case being considered, there will be essentially no variation with concentration of hydrogen ions since the law of mass action applied to the reaction

$$H_3O^+(aq) + OH^-(aq) \rightarrow 2H_2O(\ell)$$

requires that $a_{H_3O^+}\, a_{OH^-} = K_w a^2_{H_2O}$ where K_w is a constant under any specified conditions of pressure and temperature. This means that unless a_{H_2O}—the activity of water in the solution—varies, any change in the activity of hydrogen ions is likely to be exactly counterbalanced by a change in the activity of hydroxyl ions. The activity of water will differ only appreciably from unity for solutions having a high concentration of ions, unless pressure and temperature are varied considerably from 1 atm and 25 °C. The dependence of the emf on this quantity is given by

$$E = E^{\ominus} + \frac{RT}{2F} \ln (P_{H_2}/\text{atm}) + \frac{RT}{4F} \ln (P_{O_2}/\text{atm}) - \frac{RT}{2F} \ln a_{H_2O} - \frac{RT}{4F} \ln K_w$$

The effect of the changing water activity on the emf, as the concentration of aqueous potassium hydroxide increases at 25 °C, is shown in table 3.3. The pressures of hydrogen and oxygen are assumed to have the standard value here.

The emf of the hydrogen–oxygen cell will clearly depend on the partial pressures of hydrogen and oxygen supplied to the electrodes, and figure 3.1 shows this dependence graphically.

Similar expressions relating emf to concentration of reactant could be derived for all other possible fuel cell reactions, and since we have said that the actual working potential difference is usually related to the emf the importance of such relations can be seen. A further example worth quoting is for the methanol oxidation reaction

$$2CH_3OH(aq) + 3O_2(g) \rightarrow 2CO_2(g) + 4H_2O(\ell)$$

where the methanol fuel is dissolved in the electrolyte (see chapter 6). The equation relating the emf with the various pressures and concentrations is

Table 3.3 Effect of changing water activity on the emf of a hydrogen–oxygen cell as the concentration of the electrolyte, aqueous potassium hydroxide, changes

KOH concentration		Standard emf
c/per cent by mass	m/mol kg^{-1}	$E^{\ominus}$/V
1	0.18	1.229
10	1.8	1.230
20	3.6	1.232
30	5.4	1.235
40	7.2	1.243
50	8.9	1.251

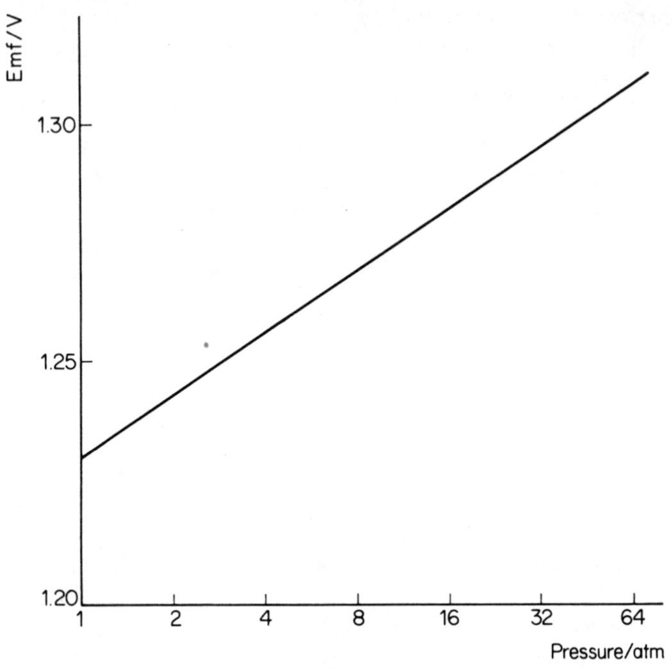

Figure 3.1 Dependence of the emf of the hydrogen–oxygen fuel cell

$$H_2(g) + \tfrac{1}{2}O_2(g) - H_2O(\ell)$$

at 298 K on pressure. The pressure axis is a logarithmic scale

$$E = E^{\ominus} + \frac{RT}{4F}\ln(P_{O_2}/\text{atm}) + \frac{RT}{6F}\ln(c_{CH_3OH}/\text{mol dm}^{-3}) - \frac{RT}{3F}\ln a_{H_2O}$$
$$- \frac{RT}{6F}\ln(P_{CO_2}/\text{atm})$$

This expression assumes no appreciable ionisation of any of the constituents and also that the gases and the methanol solution behave ideally. These assumptions are not precisely correct, of course, but will not make any substantial change to the form of the relation; this indicates that the maximum potential difference obtained from the methanol fuel cell should depend on the concentration of methanol, the water activity, the pressure of oxygen supplied and the pressure of carbon dioxide escaping from the cell.

All equations obtained in this way will have very similar form; the relation between the actual working potential difference and the emf computed from this sort of expression will depend on the polarisation effects determined by the rates of the various chemical and electrochemical processes. The nature of this dependence is described in the next chapter.

CHAPTER 4

THE RATES OF ELECTRODE PROCESSES

The thermodynamics of galvanic cells as considered in the two previous chapters can, in general, only tell us of the feasibility of certain chemical reactions occurring at electrodes to give an electric potential; they can give no information about whether such processes will take place at a measurable rate. In other words, the characteristics of working under load of any kind of electrochemical cell (including a fuel cell) cannot be deduced from computation of the changes in thermodynamic parameters associated with the cell reaction.

By definition, the electromotive force of a cell is the potential difference between its electrodes, measured when no current is flowing through the external circuit connected to the electrodes of the cell. In practice this must mean that the current flowing is too small to be detectable.

Although during potentiometric measurement of emf a minimum current flows in the external circuit, this does not imply that only the same small current flows through each part of the cell itself. The equilibrium set up between an electrode and an electrolyte must be a dynamic one; that is to say the ions passing into solution from the electrode are balanced by the ions deposited from the electrolyte onto the electrode. For example

$$Cu(s) \rightleftharpoons Cu^{2+}(aq) + 2e^{-}$$

these processes which are likely to involve the movement of ions through the solution to or from the other electrode, give rise to the phenomenon known as *exchange current:* the current carried by each of these balanced reactions between solid electrode and solvated ions. Clearly the net current will be zero, but nevertheless this results from the cancelling of two electron transfers in opposite directions. This phenomenon will be discussed in more detail in a subsequent section, but it can be seen that even at equilibrium electron transfers occur in the vicinity of the electrode, despite no overall charge transfer taking place in the cell and no external current flowing.

A working cell, that is to say one connected to an appropriate load such as a motor, an arrangement of lights or a heating system, clearly cannot possibly fulfil the conditions of thermodynamic reversibility. This does not mean that thermodynamic calculations and argument are of no value; as mentioned previously, they do indicate the feasibility of the processes involved and enable us to calculate the maximum potential difference that could be obtained for the various concentrations of material in the system. Moreover, it is frequently convenient to relate the behaviour of a working cell (where a finite current is passing) to the behaviour of the reversible system, the actual potential difference measured being regarded as a variation from the reversible value, often called 'overvoltage' or 'overpotential'.

The *overpotential* of a particular electrode system is thus defined as the difference between the measured potential under working conditions and the reversible or thermodynamic potential. For a working galvanic cell—one where a chemical reaction produces electric current—the measured potential will be lower than the reversible one, while in electrolysis—where a chemical reaction is induced by the passage of an electric current—the measured potential will be the higher of the two. Clearly, there is no fundamental difference between these systems: only the direction of the electric current in the external circuit is different, provided that the materials of the cell are identical. For electrolysis to take place it is necessary to provide an extra potential difference to overcome various energy barriers connected with the discharge of the ions and the movement of the ions through the cell electrolyte. When a galvanic cell produces current, the potential difference across it is lowered because of the energy required once again to move the ions or to surmount physical or chemical energy barriers for ionisation. In both cases these losses produce the observed overpotential.

Although the effect of polarisation at both fuel and oxidant electrode will affect the observed potential differences of the cell, it is essential for the understanding of the origins of overpotential to consider each electrode as a separate reaction site. The total, or cell, overpotential will be a combination of the individual electrode overpotentials.

Types of polarisation

As we have already hinted at in the previous paragraph, overpotential may be subdivided according to the phenomenon believed to be responsible. It is true that this may not always be realistic since the phenomena concerned may be cooperative and division may not be justified, but on the whole it is a very useful way of examining both qualitatively and quantitatively the various influences of the physical and chemical processes occurring within the cell. These processes which give rise to overpotential either in the galvanic or the electrolytic cell are often referred to as *polarisation*; in a sense they tend to set up a potential difference in opposition to that imposed by the external circuit or produced by the overall chemical reaction (as the case may be). Consequently, we divide overpotential into three types: *activation*, *concentration*, and *ohmic overpotentials*.

Activation overpotential arises both from the chemical reaction and the physicochemical processes associated with the adsorption of molecules or atoms on surfaces at the electrodes. As its name suggests, it is regarded as being directly analogous to the activation energy of the rate determining process or processes. In an ordinary chemical reaction which occurs in several stages, the rate determining step is the stage with the lowest rate of reaction on which all the other rates must depend. Sometimes there may be more than one stage of comparable slowness, or the slowest stage may be different at different temperatures. The activation energy is the energy that the reacting species (molecules, atoms or ions) need to possess before reaction to form products can actually take place. Figure 4.1 shows this idea in terms of the change of Gibbs function in the system as the reaction at a single electrode proceeds. The horizontal axis in the figure represents 'reaction coordinate' which is of course related to time elapsed (for simple reactions, this axis also represents degree of change from reactant to product). The maximum of the curve corresponds to the 'activated complex' or 'transition state' when the reactant molecules have gained their energy of activation, and have undergone partial reaction.

It can be seen that the algebraic difference of the Gibbs functions of activation ($\Delta G_1^\ddagger$ and $\Delta G_2^\ddagger$) for forward and backward reactions is equal to the change in Gibbs function for the overall reaction. This diagram assumes that the forward and backward processes follow the same reaction path, which for elementary processes and the kind of reactions we are considering is likely to be the case. Later in this chapter some of the reactions

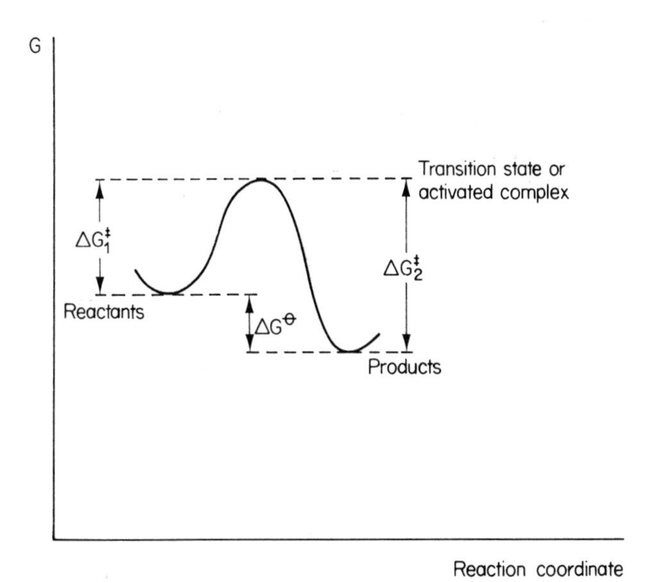

Figure 4.1 Variation of Gibbs function with reaction coordinate for a simple electrochemical reaction

taking place at fuel cell electrodes will be examined and their influence on activation overpotential considered.

Concentration overpotential arises because in most working galvanic cells the concentration of the ions which react at the electrodes will be lower in the immediate vicinity of the particular electrode than in the bulk of the solution. Clearly, other things being equal, the higher the rate of discharge of ions the greater will be this overpotential. The nature of the relation between concentration of ions in the electrolyte and the distance from the electrode where these ions are being discharged is shown in figure 4.2.

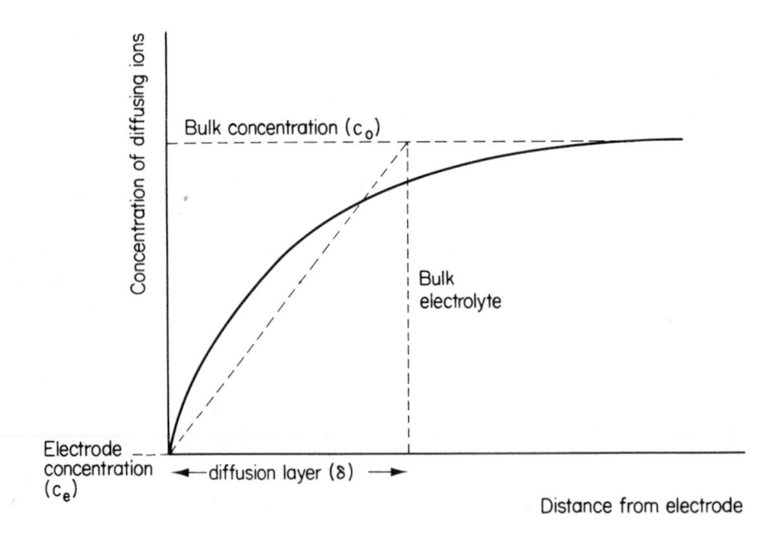

Figure 4.2 Concentration of ions being discharged at an electrode as a function of distance from the electrode. The solid line represents the actual variation of concentration, while the broken line represents the variation used in the usual treatment of diffusion overpotential

This kind of polarisation can often be reduced by some form of stirring or agitation of the cell electrolyte. Its magnitude depends on the rate of diffusion of the particular ions: this dependence is discussed in a later section.

Ohmic overpotential is simply a consequence of the electrical resistance of the solution. Clearly the greater this internal resistance the more energy will be used up in overcoming it and the lower will be the actual potential difference of the cell. For this reason it is obviously important to select as the cell electrolyte a solution which has as high a conductivity of those species reacting at the electrodes as possible.

A diagrammatic representation of the situation that is believed to exist at and around a negatively charged electrode surface is shown in the upper part of figure 4.3. The lower part of the figure indicates the attribution to the various layers (regarded as distinct in this simplified scheme) of the

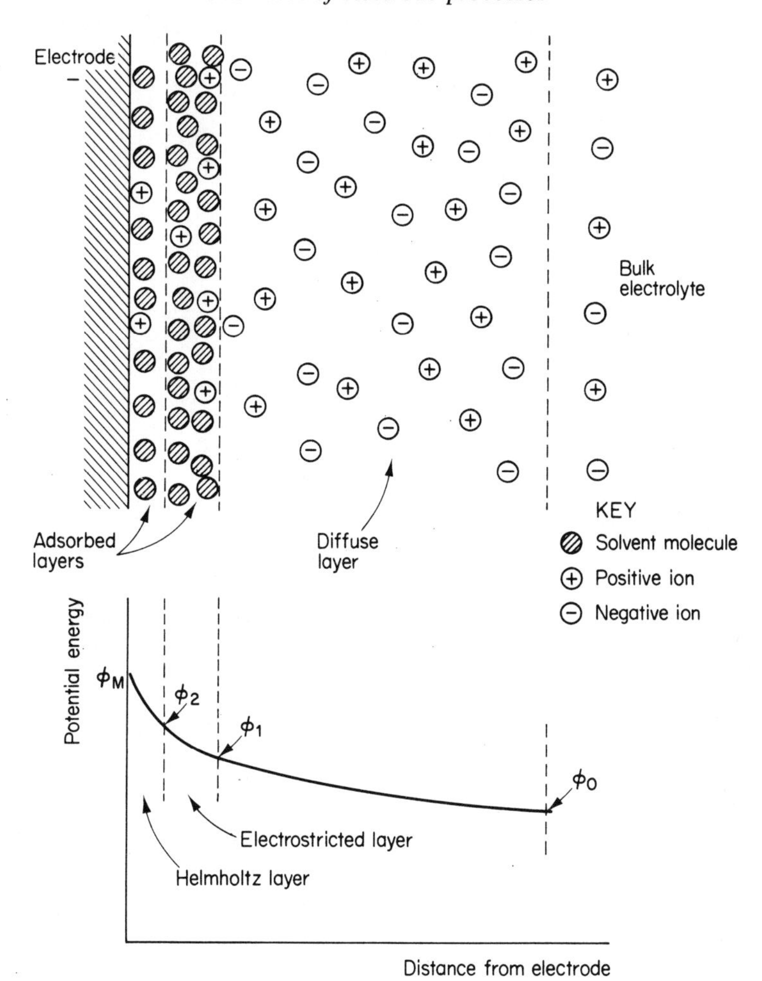

Figure 4.3 Diagram of the double layer close to a negative electrode. The potential energy of positive ions in this region when no current is flowing is shown in the lower diagram. $\phi_\mu - \phi_2$ is the electron transfer potential; $\phi_2 - \phi_1$ is related to the activation overpotential; and $\phi_1 - \phi_0$ is related to the diffusion overpotential

different kinds of overpotential and their contribution to the potential energy of the positive ions.

Activation polarisation—the role of electron transfer

The first case to be considered is that where electron transfer is the rate determining process, which may be represented by the chemical equation

$$S_{ads} \rightleftarrows S^+ + e^-$$

This can best be examined by consideration of the potential difference–

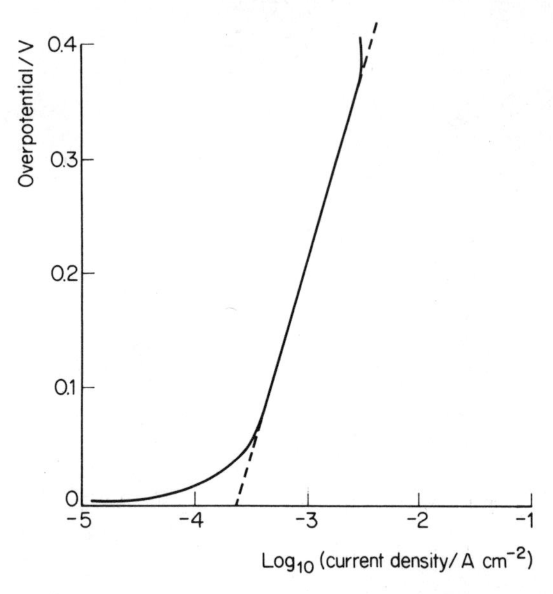

Figure 4.4 Tafel plot (for the equation $V = a \log_{10}(bi)$) for hydrogen ionisation at a solid nickel electrode in 5 mol dm^{-3} aqueous KOH at 30 atm and 150 °C

current relation. The parallel drawn between activation overpotential and chemical activation energy can be extended further by consideration of the Tafel equation. This relationship was established by Tafel in 1905, entirely on empirical grounds, for all kinds of overpotential; it can be written in two ways[1]

$$V = a \log (bi)$$

or

$$i = Be^{-AV}$$

where V is the overpotential usually at one electrode for a current density i, and a, b, A and B are empirical constants. A typical Tafel plot for a cathodic reaction (in which positive ions are discharged) is shown in figure 4.4 from which it will be seen that the equation is obeyed accurately over a substantial portion of the graph. The linear portion of the graph may be extrapolated to zero overpotential to yield a quantity i_0, the *exchange current density*. This is related to the exchange current mentioned on p. 31 and can be regarded as the current passing per unit area at an electrode within the cell (in both directions) when the external circuit is 'open', that is when the equilibrium or reversible potential (the emf) is established.

According to the transition state theory of reaction rates, the rate constant of a single stage unimolecular chemical reaction, k_1, is given by

$$k_1 = \frac{kT}{h} e^{-\Delta G_1^{\ddagger}/RT}$$

where k is the Boltzmann constant, h is the Planck constant, R is the gas constant, T is the thermodynamic temperature, and $\Delta G_1^{\ddagger}$ the free energy of activation (or Gibbs function change for the activation process) for the reaction concerned. There is a slightly different expression for reactions other than unimolecular, but most of the reactions considered may be assumed to be of the simplest chemical type.

From our earlier diagram of the Gibbs function changes during an electrode reaction (figure 4.1) we can see that a similar equation can be written for the reverse reaction

$$k_2 = \frac{kT}{h} e^{-\Delta G_2^{\ddagger}/RT}$$

where subscripts 1 and 2 refer to the forward (ionisation) and backward (discharge of ions) reactions respectively. If this reaction is a reversible process (in the thermodynamic sense) then at equilibrium the forward and backward rates will be equal, and we may write

$$k_1 \, c_{react} = k_2 \, c_{prod}$$

where c_{react} and c_{prod} are the surface concentrations of reactants and products respectively. Thus

$$\frac{c_{prod}}{c_{react}} = \frac{e^{-\Delta G_1^{\ddagger}/RT}}{e^{\Delta G_2^{\ddagger}/RT}} = e^{-\Delta G/RT}$$

This, of course, conforms with the expression for the equilibrium constant, $K = c_{prod}/c_{react}$.

If an electric potential exists between the reactants and the products—that is to say, if the reaction is one of ions dissolving at an electrode leading to charge separation between electrode and solution which is not balanced by a corresponding reaction in the reverse direction—then figure 4.1 must be redrawn. The reactants and products now have Gibbs function values different from the reversible ones since they are affected by the overpotential between them. The new diagram is compared with the old in figure 4.5 and from it can be seen that the overpotential helps the forward reaction by raising the energy of the reactants while hindering the backward reaction by decreasing the energy of the products. If we assume once again that the reaction is the dissolution of metal from an electrode to form metal ions in solution, then we may say that the overpotential increases the ease (that is the rate) of dissolution from the electrode but also decreases the rate of discharge of the ions from the solution.

It is usually assumed that the effect of the overpotential is not the same in both cases, but that only a proportion, α, sometimes referred to as the

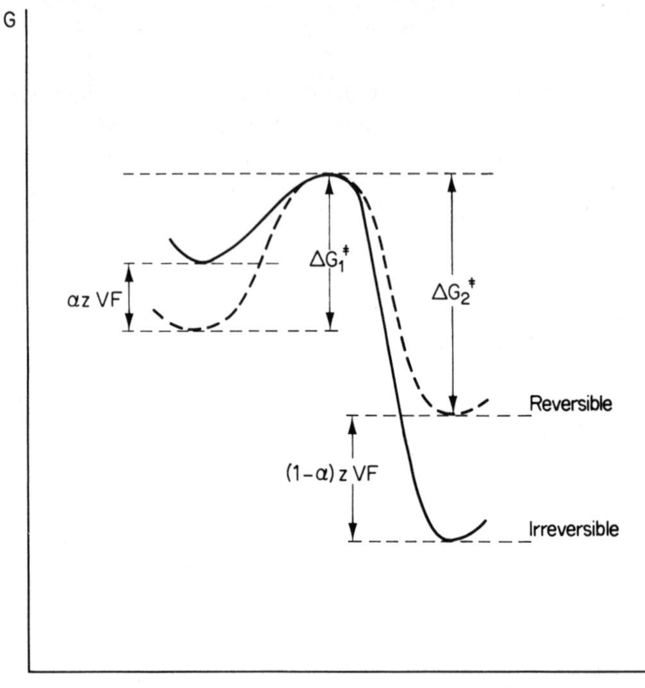

Figure 4.5 The effect of the imposition of an overpotential V on the variation of Gibbs function with reaction coordinate for a simple electrochemical reaction. The 'reversible' curve is shown also (as a broken line) for purposes of comparison

transfer coefficient, modifies the Gibbs function of the reactants, leaving a proportion $(1-\alpha)$ to affect the Gibbs function of the products. Values of the transfer coefficient for some electrode reactions are given in table 4.1. Since thermodynamic argument in chapter 3 led us to the relation

$$\Delta G = \frac{\partial G}{\partial x} = -zEF$$

(where ΔG is conventionally referred to as the Gibbs function change for the reaction but is really $\partial G/\partial x$, the rate of change of the Gibbs function of the system with extent of reaction) then we may put the change in Gibbs function arising from an overpotential V equal to $-zVF$ which means that the two energies of activation (for forward and reverse reactions) will become $\Delta G_1^{\ddagger} - z\alpha VF$ and $\Delta G_2^{\ddagger} + z(1-\alpha)VF$. Thus the rate constants, modified by the existence of activation polarisation, k_1' and k_2', will become

$$k_1' = \frac{kT}{h} \exp \frac{(-\Delta G_1^{\ddagger} + z\alpha VF)}{RT}$$

Table 4.1 Some values of the transfer coefficient, α, for reactions at a negative electrode

Reaction	Electrode	α
$Fe^{3+} + e^- \rightarrow Fe^{2+}$	Pt	0.58
$Ce^{4+} + e^- \rightarrow Ce^{3+}$	Pt	0.75
$Ti^{4+} + e^- \rightarrow Ti^{3+}$	Hg	0.42
$2H^+ + 2e^- \rightarrow H_2$	Hg	0.50
$2H^+ + 2e^- \rightarrow H_2$	Ni	0.58
$Ag^+ + e^- \rightarrow Ag$	Ag	0.55

and

$$k'_2 = \frac{kT}{h} \exp \frac{(-\Delta G_2^{\ddagger} - z(1-\alpha)VF)}{RT}$$

Expressions for the rates of these reactions per unit surface area of electrode can be obtained by multiplying the rate constant by the concentration of the reacting species, and by dividing the result by the surface area of the electrode. These rates, r_1 and r_2, are then given by

$$r_1 = K_1 \exp \frac{(-\Delta G_1^{\ddagger} + z\alpha VF)}{RT}$$

and

$$r_2 = K_2 \exp \frac{(-\Delta G_2^{\ddagger} - z(1-\alpha)VF)}{RT}$$

where K_1 and K_2 include not only the kT/h term of the earlier equations but also the surface concentration of the appropriate species. The 'reversible' activation energy terms containing $\Delta G_1^{\ddagger}$ and $\Delta G_2^{\ddagger}$ can also be combined with these quantities to give new constants K'_1 and K'_2 for particular reactions taking place on particular surfaces at particular temperatures when the concentration of the various species does not change: this will give

$$r_1 = K'_1 \exp \frac{z\alpha VF}{RT} \tag{4.1}$$

and

$$r_2 = K'_2 \exp \frac{-(1-\alpha)zVF}{RT} \tag{4.2}$$

where $K'_1 = K_1 \exp(-\Delta G_1^{\ddagger}/RT)$ and $K'_2 = K_2 \exp(-\Delta G_2^{\ddagger}/RT)$. The rates of reaction per unit area are proportional to the current densities. In the equations 4.1 and 4.2, when $V = 0$ (that is, zero overpotential, or reversible conditions), then $r_1 = K'_1$ and $r_2 = K'_2$. Thus K'_1 and K'_2 are also current densities and must be similarly proportional to the exchange current density,

i_0, mentioned earlier. We may now replace r_1 and r_2 by i_1 and i_2:

$$i_1 = i_0 \exp \frac{z\alpha VF}{RT} \text{ and } i_2 = i_0 \exp \frac{-(1-\alpha)zVF}{RT}$$

and the overall current density must be given by

$$i = i_1 - i_2 = i_0 \exp \frac{z\alpha VF}{RT} - i_0 \exp \frac{-(1-\alpha)zVF}{RT} \tag{4.3}$$

This equation, which is sometimes called the Butler-Vollmer equation, represents the general relation between the current at a particular electrode and the activation overpotential of the process taking place at that electrode. It clearly has the same form as the experimentally based Tafel equation. Certain special cases usually considered for discharge of cations lead to particular simplifications.

(1) *When the overpotential is small* ($V < 0.02$ V). The approximation $\exp(x) = 1 + x$ can be used in equation 4.3 which becomes

$$i = i_0 + i_0 z\alpha VF/RT - i_0 + i_0(1-\alpha)zVF/RT$$
$$= i_0 zVF/RT$$

Thus, in these conditions the current density is linearly related to the overpotential.

(2) *When the overpotential is large* ($V > 0.05$ V). In these circumstances the reverse reaction has a rate negligible compared with that of the forward reaction. Hence equation 4.3 becomes

$$i = i_0 \exp(\alpha zVF/RT)$$

which may be rearranged to

$$\ln\left(\frac{i}{i_0}\right) = \frac{\alpha zVF}{RT}$$

or

$$V = \frac{RT}{\alpha zF} \ln\left(\frac{i}{i_0}\right) \tag{4.4}$$

which is clearly the *Tafel equation*, mentioned earlier and written there as

$$V = a \log(bi)$$

whence we may establish that the constants a and b are given by

$$a = \frac{2.303RT}{\alpha zF} \quad \text{and} \quad b = \frac{1}{i_0}$$

and are clearly dependent on the particular electrode reaction. The exchange current density, i_0, can be readily evaluated from a graph of experimental results. Values of i_0 for hydrogen discharge can vary from about 10^{-12} A cm^{-2} at a mercury cathode, to about 10^{-3} A cm^{-2} at a platinum cathode.

If the exchange current density is less than about 10^{-7} A cm^{-2} it is unlikely that the equilibrium (or reversible) potential for this process can be established, since impurities likely to be present in the electrode material will probably yield higher exchange current densities and thus any observed potential will be due to them. On the other hand it is interesting to note that an electrode process involving a single electron transfer and no structural rearrangement, will generally have a high exchange current density—of the order of 10^{-3} A cm^{-2}. It should be mentioned here that for reasons of practical convenience current density is always defined in terms of geometrical area of electrodes rather than the actual area. Clearly, measurement of the actual area of a highly porous electrode surface would be very difficult. In some cases the ratio of actual area to geometrical area (sometimes called the *roughness factor*) may be very great; for example, about 2000 for platinised platinum.

Surface chemistry reactions

The rate controlling stage in the reaction at an electrode may not be the process of electron transfer but may be connected with the adsorption of species on the electrocatalytic surface of the electrode, or their reaction there. This situation is more likely to be appropriate in the case of hydrocarbon fuels than for hydrogen.

There are generally considered to be two steps in the surface process: adsorption on a surface site, and ionisation at that site. These may be represented by

$$\mathrm{F(g)} + \text{site} \underset{k_{-1}}{\overset{k_1}{\longrightarrow}} \mathrm{F_{ads}}$$

$$\mathrm{F_{ads}} \underset{k_{-2}}{\overset{k_2}{\longrightarrow}} \mathrm{F^+(aq)} + \mathrm{e^-}$$

(for a fuel substrate, F). k_1, k_{-1}, k_2 and k_{-2} represent the rate constants of the chemical reactions in the directions shown.

It is usual to assume that the Langmuir model adequately describes the behaviour of the adsorbed species. This means that all sites are equally reactive, whatever the surface coverage, and that adsorbed species can cover a surface until a complete unimolecular layer ('monolayer') has been formed but after that no further adsorption can occur. If θ is the fraction of surface sites covered at any particular time, then

$$\frac{\mathrm{d}\theta}{\mathrm{d}t} = k_1 P(1-\theta) - k_{-1}\theta - k_2\theta$$

where P is the partial pressure of the gaseous species F and the reverse electrochemical reaction is considered negligibly slow (that is, $k_{-2} \ll k_2$). This rate of change of surface coverage, $d\theta/dt$, will be zero in the steady state, when a current density i is flowing. Thus

$$k_1P(1-\theta) - k_{-1}\theta - k_2\theta = 0$$

and

$$\theta = \frac{k_1P}{k_1P + k_{-1} + k_2}$$

The current density flowing will be given by

$$i = zFk_2\theta c$$

(where c is the monolayer concentration) which on substitution for θ yields

$$i = \frac{zFk_1k_2Pc}{k_1P + k_{-1} + k_2} \tag{4.5}$$

Equation 4.1 can be rewritten for this ionisation reaction as

$$k_2 = B\exp(\alpha zFV/RT) \tag{4.6}$$

V being the activation overpotential, B a constant, and the other symbols having their usual meanings. Elimination of k_2 between equations 4.5 and 4.6 followed by rearrangement yields

$$V = \frac{RT}{\alpha zF}\ln\left(\frac{ik_1P + ik_{-1}}{B(PzFk_1c - i)}\right) \tag{4.7}$$

Comparison of equation 4.7 with the form of the Tafel equation derived for an electron transfer controlled process (equation 4.4) suggests that if adsorption is the rate determining step then the overpotential will increase less rapidly with increasing current than if electron transfer is rate determining.

The Temkin model is sometimes thought to be closer to the real behaviour of electrochemical surface reactions; here the Gibbs function change of adsorption is taken as proportional to the fraction of surface covered.

Table 4.2 Some values of the constants in the Tafel equation $V = a\log(bi)$ for various electrochemical processes. For convenience, $1/b$, (i_0) the exchange current density, is tabulated

Electrode	Electrolyte	Process	a/V	i_0/mA cm^{-2}
Cu	1 mol dm^{-3} $CuSO_4$(aq)	deposition of Cu	−0.051	2×10^{-2}
Ni	1 mol dm^{-3} $NiSO_4$(aq)	deposition of Ni	−0.051	2×10^{-6}
Pt	0.05 mol dm^{-3} H_2SO_4(aq)	evolution of O_2	−0.044	2×10^{-7}
Pt	1 mol dm^{-3} NaN_3(aq)	evolution of N_2	−0.026	10^{-73}
Hg	0.1 mol dm^{-3} KOH(aq)	evolution of H_2	−0.093	4×10^{-12}
Ag	7 mol dm^{-3} HCl(aq)	evolution of H_2	−0.090	1.3×10^{-3}
Pd	0.5 mol dm^{-3} H_2SO_4(aq)	evolution of H_2	−0.080	1

However, when the coverage is very low or very high ($\theta \to 0$ or $\theta \to 1$) then the equation obtained reduces to the Langmuir form.

Thus the slope (earlier represented by the symbol *a*) of the linear portion of the experimental Tafel plot may be used to distinguish between the various possible rate determining steps in the mechanism of the electrode process. For more complicated cases where neither the chemisorption nor the electron transfer stages are rate controlling, but the important stage is a surface or other reaction of the ion formed—for example a chemical rearrangement—the Tafel slope again may be significant. Some representative values of Tafel slopes for various electrochemical processes are shown in table 4.2. Further comment on this point will be found in the discussion of the behaviour of the hydrogen electrode later in this chapter.

Concentration polarisation

During the derivation of the expressions relating current to activation overpotential, in the last two sections, we made the assumption (explicitly or otherwise) that the concentration of electrolyte involved was that of the bulk solution. However, as we have already stated, the concentration of ions at the electrode surface is likely to be very different, because of the problems of mass transfer either of ions formed moving away from the electrode surface, ions to be discharged moving to the electrode, or of other species moving to the electrode to react with ions formed there. In the case of both fuel and oxidant electrodes of a fuel cell, the first and third of these alternatives seem the most probable. There are two other kinds of mass transfer which may be important in particular cases. Diffusion of the reacting gas in the pores of the electrode is one of these, but this is usually only significant when there is an inert gas present (for example, nitrogen in air being fed to an oxygen electrode) or when a gas is being discharged at the electrode concerned (for example, carbon dioxide is discharged at the fuel electrode of some high temperature cells). The other kind concerns the movement of fuel molecules from their supply at the electrode/electrolyte interface to the reactive sites, generally through the electrolyte.

The most satisfactory mathematical treatment is that of substances diffusing from the bulk electrolyte to the electrode surface, where they react, presumably with ions formed there. For the sake of simplicity three regions of electrolyte are defined: a thin stagnant layer very close to the electrode surface (or more correctly close to the electrical double layer bounding the electrolyte), the 'bulk' electrolyte (the major part of the electrolyte where concentrations of all species are essentially constant), and the *diffusion layer*, lying in between the other two (see figure 4.3).

This diffusion layer is shown schematically in figure 4.2 which also shows a linear change in concentration of the species concerned across this layer. Such a variation is almost certainly unreal (the supposed real behaviour is also shown in the figure), but has again the merit of simplicity of treatment.

If the concentrations are c_0 in the bulk region and c_e at the electrode, then the rate of flow is equal to $AD\ (c_0 - c_e)/\delta$ where the cross sectional area is A, the thickness of the diffusion layer is δ, and D is the *diffusion coefficient*.

The value of the diffusion coefficient depends on the nature and size of the solute species, on the viscosity of the medium and on its temperature. It decreases with increase in viscosity and with decrease in temperature; a typical value at ordinary temperatures is about $10^{-9}\ m^2\ s^{-1}$. The rate of flow can be replaced by iA/zF where i is the current density, F the Faraday constant, and z the number of electronic charges carried per ion, and hence

$$\frac{i}{zF} = \frac{D(c_0 - c_e)}{\delta}$$

or

$$i = \frac{DzF(c_0 - c_e)}{\delta} \tag{4.8}$$

The diffusion layer thickness, δ, is taken to be about 300 μm for unstirred solutions and about 1 μm for vigorously stirred systems.

When $c_e = 0$, that is, when every ion arriving at the electrode surface immediately reacts, then the *limiting current density* is given by

$$i_L = \frac{DzFc_0}{\delta_L}$$

D is assumed not to change across the diffusion layer, and it is found to vary little with concentration. Equation 4.8 may be rewritten

$$\frac{i}{i_L} = \frac{c_0 - c_e}{c_0}\frac{\delta_L}{\delta}$$

If we assume that the thickness of the diffusion layer does not change much with concentration (which may be true), then $\delta_L = \delta$ and we obtain

$$i = (1 - c_e/c_0)i_L \tag{4.9}$$

By regarding the change of concentration across the diffusion layer as producing an overpotential V we may, by assuming thermodynamic behaviour and the type of equation obtained in chapter 3 for dependence of electrode potential on ionic concentration, write

$$V = \frac{RT}{zF}\ln\frac{c_0}{c_e}$$

(Of course the system is in no sense at equilibrium but there is some excuse for applying this sort of argument to a steady state condition. Full details of the justification cannot be given here but readers are referred to any general account of electrochemical kinetic phenomena—see the Bibliography.

Using the result of equation 4.9 we have

$$V = \frac{RT}{zF} \ln\left(\frac{i_L}{i_L - i}\right)$$

thus giving the overpotential–current relation for mass transfer by diffusion of ions only.

If ionic transport (that is, movement due to an applied electric field) is also a contributor to the concentration of ions at the electrode surface, then the relations obtained so far must be modified somewhat. If the movement of negative ions to the electrode is concerned as, for example, for the process

$$\tfrac{1}{2}H_2 + OH^- \rightarrow H_2O + e^-$$

then the current density is equal to the sum of two types of mass transfer—the diffusion contribution $DzF(c_0 - c_e)/\delta$ and the ionic transport contribution $t_- i$, where t_- is the transport number of the anion (hydroxide in the example).

Thus

$$i = \frac{zDF(c_0 - c_e)}{\delta} + t_- i$$

Table 4.3 Some values of conductivity for certain electrolytes

Electrolyte	Temperature/°C	Concentration/mol dm^{-3}	Conductivity/S cm^{-1}
H_2SO_4 in water	18	5	1.35
		10	1.41
		30	0.190
NaOH in water	18	5	0.345
		10	0.205
		15	0.110
KOH in water	18	5	0.528
		10	0.393
NaOH in water	50	5	0.670
		10	0.575
		15	0.440
	100	5	1.24
		10	1.41
		15	1.33
H_3PO_4 in water	18	6	0.625
		11	0.151
NaCl, fused	750	—	3.40
Na_2CO_3, fused	850	—	2.92
KCl, fused	900	—	2.76

or

$$i = \frac{zDF(c_0 - c_e)}{\delta(1 - t_-)}$$

which may be rewritten as

$$i = zDF(c_0 - c_e)/t_+ \delta$$

where t_+ is the transport number of the *cation*. Consequently, the limiting current density is given by

$$i_L = \frac{zDFc_0}{t_+\delta}$$

Ohmic polarisation

The overpotential, V, arising from the resistance, R, of the electrolyte may be simply expressed as a function of the current density i

$$V = iAR$$

where A is the cross sectional area of the electrodes. The resistance is related to the *conductivity*, k (formerly known as specific conductivity), to the distance, x, separating the electrodes, and to the cross sectional area A:

$$R = \frac{x}{kA}$$

Hence

$$V = \frac{ix}{k}$$

For a current density of 200 mA cm^{-2}, a cell with electrodes 2 mm apart and an electrolyte having a conductivity of 0.4 S cm^{-1}, there will be an ohmic overpotential of about 0.1 V. Since the potential of most fuel cells is of the order of 1 V and since the figures chosen are quite representative, we can see that ohmic polarisation can have an appreciable effect on working potentials. Some figures for conductivity of selected electrolytes are shown in table 4.3. Some minor increases in calculated overpotentials may arise from the changes in resistance of the electrolyte because of the removal, production or replacement of ions at the electrodes. A most important example of such a case is the production of carbon dioxide from hydrocarbon fuel oxidation and its reaction in alkaline electrolytes to replace hydroxide ion by the slower moving (and hence less conducting) bicarbonate ion.

The hydrogen electrode

We can now consider the kinetic processes which seem to be important in a representative fuel electrode, namely the hydrogen electrode. We have

Table 4.4 Values of the Tafel slope (a) for hydrogen evolution at various electrodes from various aqueous solution. The exchange current density ($i_0 = 1/b$) is also shown

Electrode	Electrolyte	Concentration/mol dm^{-3}	$-\log_{10}(i_0/\text{A cm}^{-2})$	a/V
Pt	HCl	0.5	2.6	0.028
Pd	H_2SO_4	0.5	3	0.080
Cu	HCl	0.1	6	0.117
Cu	NaOH	0.15	6	0.117
Ag	HCl	1.0	4	0.130*
Ag	HCl	1.0	5	0.060†
Au	HCl	0.1	5	0.097*
Au	HCl	0.1	6	0.071†
Cd	H_2SO_4	0.85	12	0.120
Hg	HCl	1.0	12	0.119
Hg	LiOH	0.1	12	0.102
Al	H_2SO_4	1.0	10	0.100
Sn	HCl	1.0	8	0.140
Pb	HCl	1.0	13	0.119
Pb	H_2SO_4	10.0	13	0.119
Mo	HCl	0.1	6	0.104*
Mo	NaOH	0.1	7	0.116*
Mo	HCl	0.1	7	0.080†
Mo	NaOH	0.1	7	0.087†
W	HCl	5.0	5	0.110
Fe	HCl	1.0	6	0.130
Fe	NaOH	0.1	6	0.120
Ni	HCl	1.0	5	0.109
Ni	NaOH	0.1	5	0.101

* *high current density*, $i = 0.01$–01 A cm^{-2}
† Low current density, $i = 0.01$ A cm^{-2}

already seen that the exchange current density for hydrogen ion formation at an electrode may vary over quite wide limits: for example, i_0 at a platinum cathode is about 10^{-3} A cm^{-2} whereas it is about 10^{-12} A cm^{-2} at a mercury cathode. These are extremes but they illustrate the considerable influence of the electrocatalytic surface on polarisation of the hydrogen electrodes. Moreover, the actual condition of the surface, that is its pore structure and the minute traces of impurities that may be present, can also have a considerable effect on overpotential.

The most probable mechanism for the ionisation of hydrogen at electrodes of different materials is best characterised by the slope of the experimental Tafel equation, $V = a \log(bi)$; some representative values both for acid and alkaline solution are shown in table 4.4 which also gives approximate values of the exchange current density at these electrodes. From these figures it is possible to distinguish three classes of electrode behaviour.

(1) Metals with moderate strengths of adsorption and high values of i_0 (for example, Pt).

(2) Metals which adsorb hydrogen weakly and have low values of i_0 (for example, Hg).

(3) Metals which adsorb hydrogen strongly and have low values of i_0 (for example, Fe).

These types of behaviour can be explained mechanistically by interpreting the activation overpotential as arising from different rate determining steps.

Type 1 metals (moderate adsorption) seem to be characterised by a slow chemisorption adsorption stage

$$H_2 + M \rightarrow 2M \ldots . H$$

Type 2 metals (weak adsorption) are characterised by a similar mechanism to type 1 or alternatively by a fast adsorption step followed by slow electron transfer

$$H + H_2O \rightarrow H_3O^+ + e^- \qquad \text{(acid solution)}$$

Type 3 metals (strong adsorption) seem to be characterised by a slow reaction between chemisorbed hydrogen molecules and water

$$M \ldots . H_2 + H_2O \rightarrow M \ldots . H + H_3O^+ + e^-$$

although the actual adsorption process may be fast or slow.

Selection of fuel electrodes for hydrogen–oxygen fuel cells clearly depends on these factors but also must depend on the thermodynamic stability of the metal chosen in the acid or alkaline solution. This generally means that the most kinetically active metals are ruled out; the platinum metals, copper, silver, gold and nickel (except in acid solution) are about the only ones left, although non-metallic materials such as carbon and nickel boride can also be used.

Investigations have shown that much the same considerations made for the electrocatalytic oxidation of hydrogen apply to hydrocarbon fuels. The order of effectiveness is similar, although there are some exceptions and specificity is observed in certain cases.

The oxygen electrode

We have already mentioned in chapter 1 that oxygen is generally preferred as the oxidant in fuel cells; it is readily obtainable from the air, easily stored and handled, and gives fairly innocuous products of reaction. Such characteristics do not apply to the other contenders, such as chlorine or fluorine, which may have superficial thermodynamic attraction.

Unfortunately a really satisfactory oxygen electrode, operating in a neutral or weakly acidic medium as well as a strongly alkaline one, has never been realised. The electrode works at its best when operating in strongly alkaline solutions (for example 40 per cent aqueous KOH at 25–60 °C), when a porous silver electrode would give a current density of 50 mA cm^{-2} at a polarisation of 0.35 V. The same electrode in a saturated aqueous solution of potassium bicarbonate (pH 8.8) would give only 1 mA cm^{-2} at this polarisation, and in a solution of pH 3.0 would give no observable current at all. Curiously, strongly acidic media are more satisfactory; a figure of 10 mA cm^{-2} at 0.55 V polarisation was obtained for a platinum activated carbon electrode in 3.5 mol dm^{-3} aqueous H_2SO_4 at 60 °C.

At higher temperatures the oxygen electrode is less troublesome and polarisation is much less significant. Further discussion of this point will be found in chapters 7 and 8.

The mechanism of operation of the oxygen electrode at lower temperatures is thought to be as follows. The first stage is a two point molecular adsorption of oxygen on two adjacent sites on the electrode surface

$$M + M + O_2 \rightarrow M \ldots\ldots O{=}O \ldots\ldots M$$

Several possibilities can now occur: one is that two electrons are accepted by the adsorbed oxygen

$$M \ldots\ldots O{=}O \ldots\ldots M + 2e^- \rightarrow M \ldots\ldots O^- {-} O^- \ldots M$$

followed by hydration with two water molecules

$$\underset{\underset{M}{\vdots}}{O^-}-\underset{\underset{M}{\vdots}}{O^-} + 2H_2O \rightarrow H-\underset{\underset{M}{\vdots}}{O}-H-\underset{\underset{M}{\vdots}}{O^-}-\underset{\underset{M}{\vdots}}{O^-}-H-\underset{\underset{M}{\vdots}}{O}-H$$

This complex then splits in two

$$H-\underset{\vdots}{O}-H-\underset{\vdots}{O^-}-\underset{\vdots}{O^-}-H-\underset{\vdots}{O}-H \rightarrow H-\underset{\vdots}{O}-H-\underset{\vdots}{O^-} + \underset{\vdots}{O^-}-H-\underset{\vdots}{O}-H$$

and each of these entities can then accept a further electron to become two hydroxyl ions (still on the surface)

$$H-\underset{\vdots}{O}-H-\underset{\vdots}{O^-} \rightarrow H-\underset{\vdots}{O^-} + H-\underset{\vdots}{O^-}$$

and finally these are desorbed

$$M \ldots\ldots O{-}H \rightarrow M + OH^-$$

Slight variation of these stages could well occur and in particular production of hydrogen peroxide, H_2O_2, or its associated ion HO_2^-, can be postulated. In fact, production of hydrogen peroxide has been observed around oxygen electrodes in certain circumstances. The overall reaction would be

$$O_2 + H_2O + 2e^- \rightarrow HO_2^- + OH^-$$

as compared with

$$O_2 + 2H_2O + 4e^- \rightarrow 4OH^-$$

for the main process. Hydrogen peroxide will in any case usually be readily decomposed by subsequent catalytic or electrochemical action:

$$H_2O_2 \rightarrow H_2O + \tfrac{1}{2}O_2$$

or

$$HO_2^- + H_2O + 2e^- \rightarrow 3OH^-$$

Electrochemical experiments have shown that any catalyst that facilitates either of these two decompositions of hydrogen peroxide, particularly the second, is likely to be a favourable choice for a working oxygen electrode; materials found satisfactory are those which can adsorb OH^- and related species (for example HO_2^-) easily and can also release them easily but will not adsorb other ions (which might block the sites). so readily. The best arrangement seems to be a porous carbon base with small amounts of platinum or silver or the oxides of nickel, cobalt or copper contained on its surfaces.

Two matters related to modification of the oxygen electrode must be considered. Since oxygen is customarily obtained from the air it is tempting to supply air directly to the electrode. In general this is found to reduce the output of the cell for two reasons: first, the air contains only about 20 per cent of oxygen and therefore higher pressures and flow rates will be needed to maintain the same current, and secondly the presence of the nitrogen and other inert gases will lead to a mass transfer problem of oxygen reaching the electrocatalytic surfaces and also of the escape of the non-reacting species. Thus an additional overpotential term will appear.

It has sometimes been suggested that the oxygen electrode could be replaced by some suitable redox system with good polarisation characteristics, such as

$$Fe^{3+} \rightarrow Fe^{2+} + e^-$$

and some arrangement for regeneration of the oxidised species with gaseous oxygen

$$4Fe^{2+} + O_2 + 2H_2O \rightarrow 4Fe^{3+} + 4OH^-$$

although this will probably have to take place somewhere separated from the fuel electrode compartment by a membrane permeable to hydrogen ions only. The chief disadvantages of this superficially attractive scheme are the additional resistance introduced by the membrane and the necessity of finding a catalyst for speeding up the regeneration reaction. Of course redox systems other than iron (III)/iron (II) could be used.

Overall performance

The working characteristics of a fuel cell will then clearly be dependent on the size of all kinds of polarisation likely to be encountered. The relative magnitudes of the various effects for an operating cell are illustrated in figure 4.6. This graph shows the various overpotential losses at different current densities for a Bacon hydrogen–oxygen cell, of the type described in chapter 7, operating at 200 °C and about 30 atm. The large oxygen electrode overpotential should be noted.

The complete potential difference–current density relation can be summarised in the equation

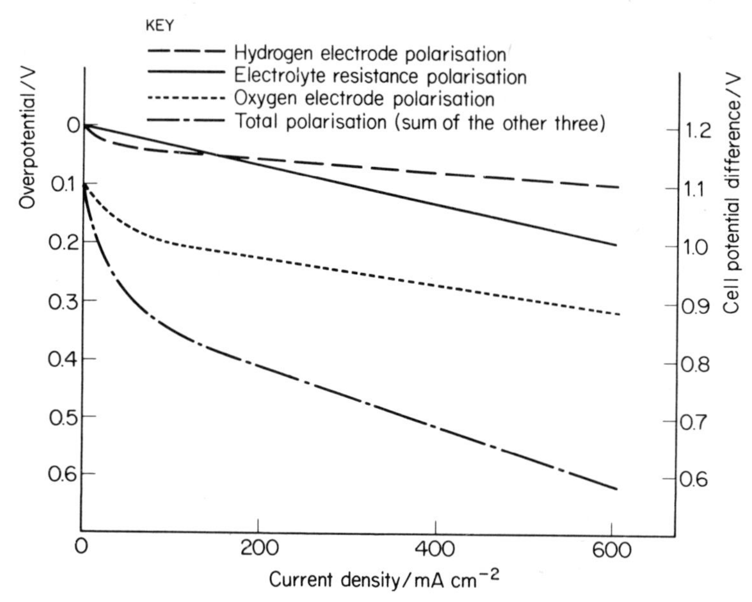

Figure 4.6 The various polarisation losses in a high pressure hydrogen–oxygen fuel cell at 473 K

$$E' = E - \frac{RT}{\alpha_C zF} \ln\left(\frac{i}{i_C}\right) + \frac{RT}{zF} \ln\left(1 - \frac{i}{i_{LC}}\right) - \frac{RT}{\alpha_A zF} \ln\left(\frac{i}{i_A}\right) + \frac{RT}{zF} \ln\left(1 - \frac{i}{i_{LA}}\right) - ri$$

where E' is the working potential difference at a current density i, E is the reversible thermodynamic emf, α_A and α_C are the transfer coefficients of anode and cathode reactions (that is, at fuel and oxidant electrodes respectively), i_C and i_A are the cathodic and anodic exchange current densities, i_{LC} and i_{LA} are the cathodic and anodic limiting current densities, z is the number of electrons transferred from cathode to anode, r is the internal resistance of the cell per unit area of electrode, and R, T and F have their usual meanings.

Over much of the current density range, this equation can be approximated:

$$E' = E_0\left(1 - \frac{i}{i'}\right)$$

where E_0 is the potential extrapolated to $i = 0$, and i' is the current density extrapolated to $E' = 0$. Figure 4.7 shows a typical plot.

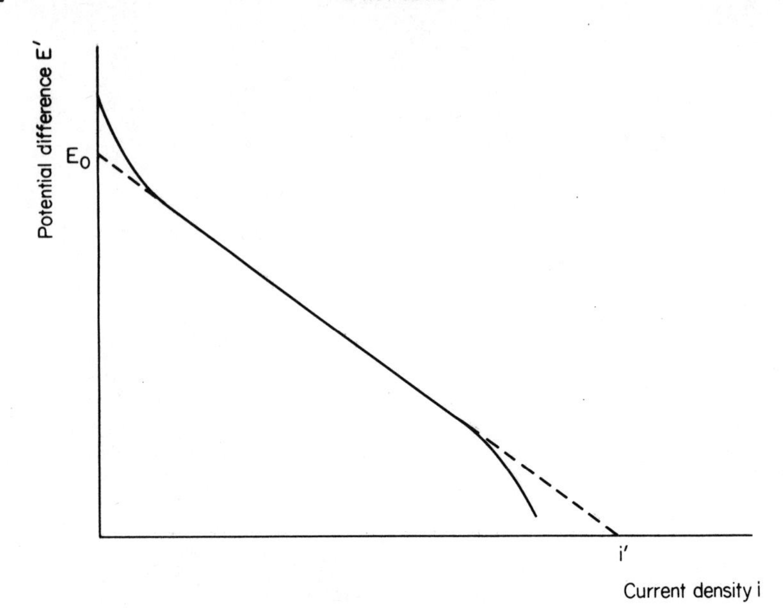

Figure 4.7 A typical plot of potential difference against current density, showing the straight line corresponding to $E' = E_0\left(1 - \frac{i}{i'}\right)$

Notes

1. The original Tafel equation was of the form

$$V = x + y \log i$$

but modern practice frowns on the use of logarithms of quantities, such as i, rather than dimensionless numbers, such as bi. Clearly, ignoring units we can write

$$x = a \log b \qquad \text{and} \qquad y = a$$

Another possible way of writing the equation in conformity with modern views would be

$$V = x + y \log (i/\text{A cm}^{-2})$$

but this is not very general.

CHAPTER 5

LOW TEMPERATURE HYDROGEN–OXYGEN CELLS

It is, in many ways, convenient to classify fuel cells according to their temperature of operation although obviously some characteristics and properties will not vary with the temperature of operation of the system. Thus in this chapter hydrogen–oxygen cells working at temperatures below about 100 °C and at low pressures (up to about 5 atm) are described, and subsequent chapters deal with other low temperature systems, and with cells working at medium (100–500 °C) or high temperatures (above 500 °C). Certain rather special types of fuel cell, such as the biochemical cell, are dealt with separately.

Limiting problems of low temperature hydrogen–oxygen cells

The significant features of technological interest in this type of fuel cell are primarily the electrodes and the electrolyte. Construction never poses a major problem for operation at low temperatures and devices to provide heat or maintain high pressures are clearly not relevant. There are likely to be problems connected with the production of water (a natural consequence of cell operation), since at low temperatures this cannot be naturally dispersed by—for example—evaporation. Peripheral equipment may therefore be required.

At low temperatures aqueous electrolytes are generally the most appropriate and their characteristics must consequently be carefully considered. The problems of electrode construction depend largely on selection of a suitable blend of catalyst materials to enable the reactions of hydrogen to give hydrated protons and oxygen to give hydroxide ions to proceed as efficiently as possible at the lowest possible overpotential. The varieties which have been used may be conveniently classified as based on metal or based on carbon. Whatever method of construction or material is employed, it is quite clear that the arrangements of pores and their sizes are particularly significant; the triple interface electrode material–gas–electrolyte must occur within these electrode pores, which means that 'flooding' either by electrolyte or by the gas is not acceptable.

Types of metal based electrodes

Certain metals are known to have high electrocatalytic activity for the gases involved; these include platinum, palladium, nickel and silver and some intermetallic compounds and alloys. Electrodes constructed of these materials are therefore likely to be appropriate for use in the low temperature hydrogen–oxygen cell.

Raney electrodes

A type of electrode worth considering in some detail is the Raney electrode: Raney metals are metals in very finely divided, highly active form. They are prepared by mixing the active metal with an inactive metal, usually aluminium. This intimate mixture (it is not a true alloy) is then treated with strong alkali which dissolves out the aluminium, leaving the highly developed surface structure desired for the porous electrodes.

The hydrogen electrode may be made from Raney nickel, usually by preparing the electrode structure from a sintered mixture of Raney alloy (50 per cent Ni 50 per cent Al) and ordinary nickel. After removal of most of the aluminium with aqueous potassium hydroxide solution, the resulting product has the necessary high activity (from the Raney nickel) and also mechanical strength (from the ordinary nickel). This electrode is highly active after treatment with hydrogen gas, and in fact is pyrophoric—that is to say it will catch fire in air when dry. Thus it is necessary to store such electrodes in water after they have been prepared.

A similar kind of electrode can be made for the oxygen side of the cell:

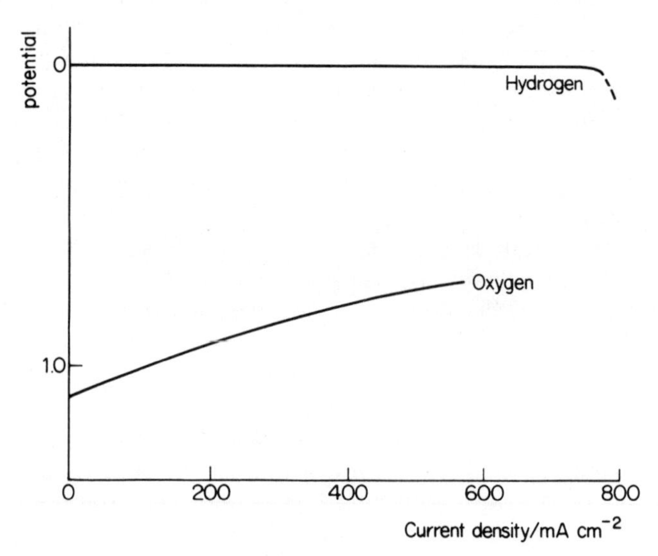

Figure 5.1 Polarisation curves for hydrogen and oxygen Raney electrodes in 6.0 mol dm^{-3} aqueous KOH at 1 atm pressure at 80 °C

in this case the mixture is of ordinary nickel and Raney silver but the surface must be prepared in a slightly different way because of metallurgical problems. In both types of electrodes the outer particles are about 6 μm in diameter and the inner layer somewhat bigger, about 6–15 μm.

Figure 5.1 shows some typical polarisation curves for these hydrogen and oxygen electrodes.

Palladium electrodes

Palladium is well known as a metal with the unusual property of allowing hydrogen to diffuse through its lattice; moreover it is electrocatalytically active for the hydrogen reaction and will not suffer from electrolyte flooding (that is, electrolyte penetrating the pores of the electrode and preventing proper gas–metal contact). It has a further useful property in that hydrogen is the *only* gas that can diffuse through its pores. Thus a palladium hydrogen electrode has many advantages and can even be used with an impure hydrogen supply, since impurities are 'filtered out'. Of course, the resulting potential will be lower because of the effect of reduced partial pressure of hydrogen (see chapter 3), and this will also affect the current that can be produced.

The electrode is made as a very thin membrane of palladium–silver alloy (which has a better mechanical strength than pure palladium) covered on both surfaces by a fine layer of palladium black. Its polarisation characteristics depend on partial pressure of hydrogen, on thickness of the membrane (the thinner the electrode the higher the current permitted), and on temperature. Figure 5.2 shows some potential current curves for a palladium–hydrogen electrode. The increase in polarisation with decrease in hydrogen partial pressure and the high cost of palladium mean that this

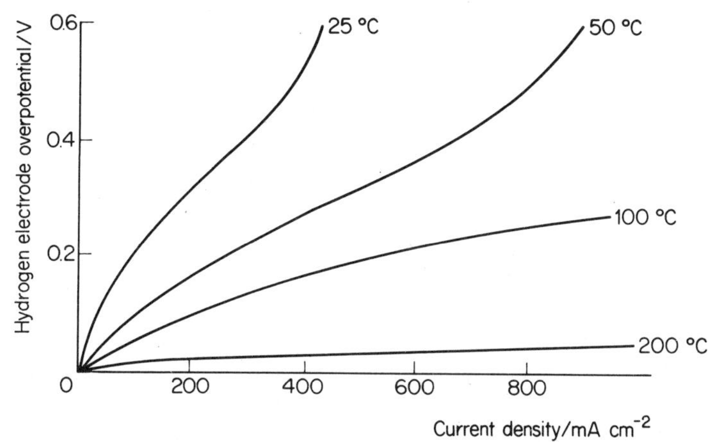

Figure 5.2 Polarisation curves at different temperatures for a palladium–hydrogen electrode

type of electrode, despite its superficial attraction, is rarely practicable in any commercial fuel cell application.

Platinum electrodes of similar design can also be used, of course, although platinum does not have such a striking property of allowing hydrogen diffusion as does palladium. But the metal is somewhat stronger mechanically and is less prone to damage by oxidation.

Nickel boride electrodes

Nickel boride is widely used as a hydrogenating catalyst in the large scale synthesis of many organic chemicals. It is not a single chemical compound, but the phase diagram of nickel and boron shows recognisable features corresponding to Ni_2B, Ni_3B_2 and NiB, and it is most conveniently prepared by precipitation from an aqueous solution of nickel chloride and sodium borohydride:

$$3NiCl_2 + 2NaBH_4 \rightarrow Ni_3B_2 + 4HCl + 2NaCl + 2H_2$$

Hydrogen electrodes are best made by depositing a layer of the nickel boride on a porous nickel structure, since this then has the necessary mechanical strength.

The catalytic effect appears to depend largely on the surface area and, more interestingly, on the amount of nickel in the sample. The second property can be shown by allowing a nickel boride electrode to stand in an aqueous

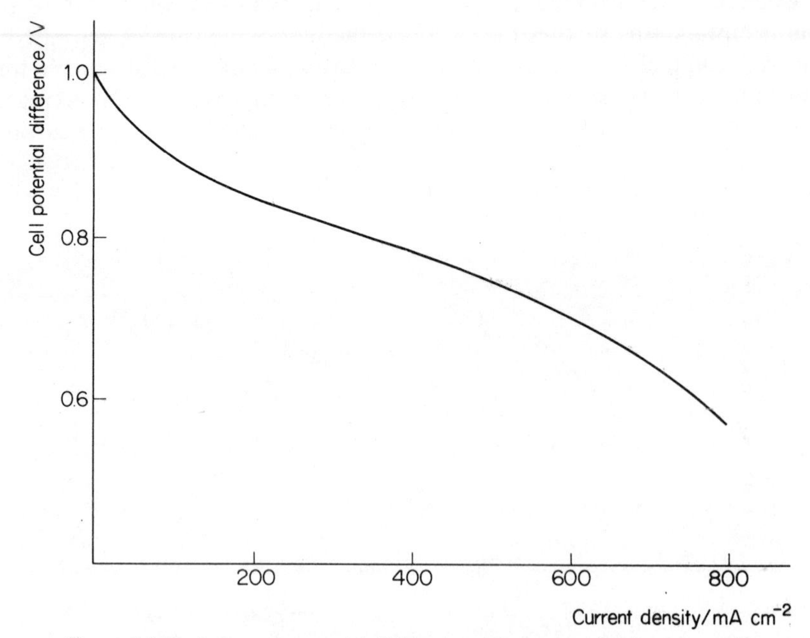

Figure 5.3 Variation of potential difference with current density for a cell with nickel boride electrodes operating at 420 K

alkaline solution (often used as an electrolyte anyway) when its overpotential for hydrogen drops with time. This seems to be due to the alkali dissolving out some of the boron, possibly via the reaction

$$Ni_3B_2 + 6OH^- \rightarrow 3Ni + 3H_2 + 2BO_3^{3-}$$

A diagram of potential against current for a cell containing this kind of electrode is shown in figure 5.3.

It is clear that nickel boride electrodes are eminently suitable as hydrogen electrodes, and may in fact be superior to the purely metallic variety, such as those made of platinum, since their cost is so much less.

Unfortunately, although nickel boride shows some electrocatalytic activity in acid electrolytes, electrodes made of this material cannot be used in such media since nickel boride is dissolved by aqueous acids.

Carbon based electrodes

Porous carbon electrodes have been used for many years in air depolarised cells (see chapter 9). Carbon itself is effective as a catalyst for oxygen electrodes in alkaline electrolytes, although its activity can be increased substantially by the incorporation of small amounts of certain active metals. Carbon is not electrocatalytic for hydrogen and consequently electrodes must be impregnated with noble metals such as platinum, iridium or palladium before they are effective.

Electrodes made for the air depolarised cell are generally of slab form but highly porous so that the air diffuses through them from one side to the electrolyte at the other—a common arrangement in fuel cells, of course. One problem with material of such variable porosity as this kind of carbon (made usually by heating suitable coals in the absence of air) is that 'flooding' by electrolyte can rather easily occur. This means that the very fine pores become full of electrolyte and hence the gas cannot gain the necessary access. It is true that the method of production generally gives a thin layer of hydrophobic material (presumably from the tar content of the coal being pyrolysed) over the surfaces, but this covering has only a short lived effect and must be enhanced by coating with some appropriate hydrophobic compound, such as paraffin wax, polyethylene or poly(tetrafluoroethylene) (PTFE), applied as a solution in some readily removable solvent.

Different forms of the carbon electrode can be used for the different shapes of battery required: for example, tubular electrodes with the gas passing through the tube and the electrolyte outside it may be suitable for some types of arrangement, whereas the plate variety may be more convenient for other styles. Diagrams of some arrangements which have been used are shown in figures 5.4 and 5.5.

There are some other variants. The high porosity carbon (which may be weak structurally) can be formed on a metal grid in the shape of the electrode; this metal grid will also serve as a current collector. A newer idea is to

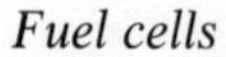

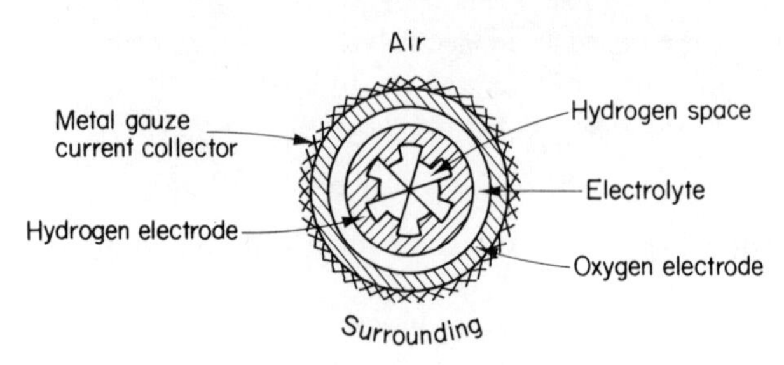

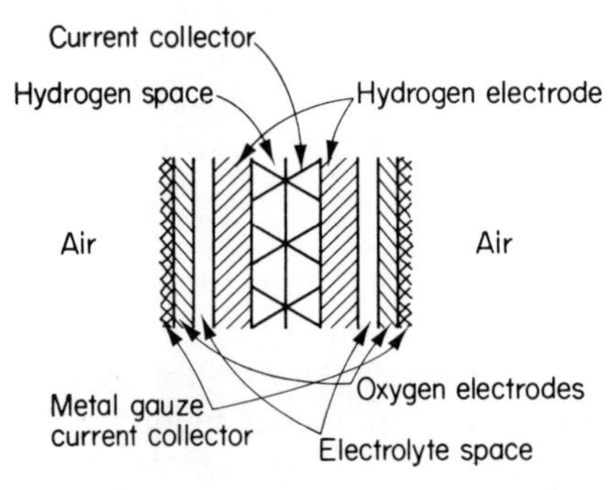

Figure 5.4 Sections of a concentric cell containing carbon electrodes

spray a very thin layer of active carbon (0.1–0.4 mm) onto a porous, hydrophobic backing material; and a further development is the construction of an oxygen electrode by depositing a thin layer of silver onto a porous plastic non-conducting material—largely poly(vinyl chloride)—about 0.8 mm thick with 5 μm diameter pores. This last variety does not of course use carbon at all.

The carbon used for electrode manufacture may be treated in other ways before installation. Heating in an atmosphere of carbon dioxide removes some carbon

$$CO_2 + C \rightleftharpoons 2CO$$

and produces various sizes of pore according to the length of time the treatment is carried out. Metals and metal oxides may be incorporated in the electrode structure by soaking the carbon in a solution of a suitable salt of the metal (or mixture of salts) before the baking process. Suitable salts will be ones which readily decompose on heating to give the metal or its oxide, whichever is required. For example the Co–Al spinel mixed oxide catalyst may be produced within a carbon structure by soaking it in a mixture of aqueous cobalt (II) nitrate and aluminium nitrate and then heating it to about 700 °C.

Carbon electrodes for the hydrogen side of the cell require impregnation with catalytic metals, as has been mentioned earlier. A method of covering

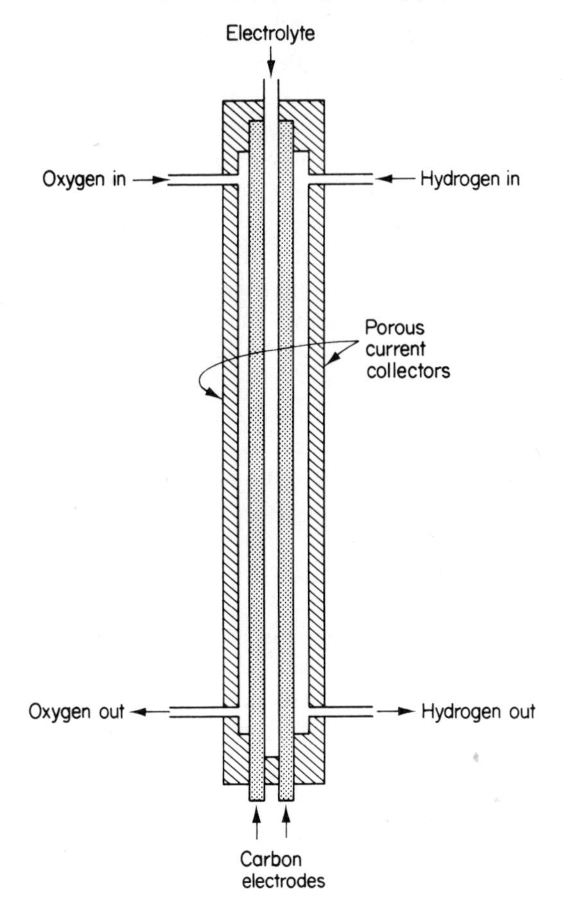

Figure 5.5 Section of a parallel plate cell containing carbon electrodes

the surfaces with finely divided platinum is to immerse an electrode already treated with the hydrophobic material in an aqueous solution of chloroplatinic acid. Heating at about 200 °C causes decomposition of this substance to the metal:

$$H_2PtCl_6 \rightarrow Pt + 2HCl + 2Cl_2$$

A relatively small amount of platinum is used in preparation of such electrodes and the cost is comparable with the Raney nickel electrode.

The life of carbon electrodes, and of the oxygen variety in particular, depends on several characteristics of the cell or battery operation. Three such characteristics are: electrolyte temperature (which will depend largely on the current density taken from the cell), removal of the water formed by the reaction, and attack of the carbon by hydrogen peroxide. Factors determining the optimum temperature and the rate of removal and production of water are discussed on p. 65 and the peroxide problem can be alleviated by use of electrodes containing peroxide decomposing materials such as the Co–Al spinel mentioned earlier.

A curiosity of the carbon oxygen electrode is that it works considerably less satisfactorily with air than with pure oxygen; this is because the high concentration of nitrogen in air (about 80 per cent) can block pore activity by its slow diffusion through the system. Nevertheless, as we indicated at the beginning of this section, this kind of electrode has actually been used, apparently successfully, in air depolarised cells.

Oxygen electrodes have also been made satisfactorily from platinum doped sodium tungsten bronzes. A sodium tungsten bronze is a non-stoichiometric compound NaW_xO_3 where $0 < x < 1$. If $x = 0.25$ then the material exhibits metallic conduction characteristics. Preparation is by cathodic deposition from a fused mixture of Na_2WO_4 and WO_3 with a platinum anode, which provides platinum for the 'doping'.

Choice of electrolyte

The choice of electrolyte for cells in the category we are considering is not very great. It is essential to use an electrolyte containing some form of solvated proton and a hydroxide ion or some closely allied species, and it is likely that any solvent other than water would probably increase the cost of any fuel cell system too considerably. Moreover, water is formed in any case by the operation of the cell, that is from the combination of H^+ and OH^- ions.

Hence we have to confine our attentions to acidic and basic substances dissolved in water. Although all aqueous solutions (whether ionic or not) must contain some H^+ and OH^- ions (for example pure water at 25 °C contains H^+ and OH^- ions each at a concentration of about 10^{-7} mol dm^{-3}) it has been shown in the previous chapter that the higher the conductivity of the electrolyte the lower the polarisation loss. Table 5.1 shows the values of limiting molar conductivity for several ions in aqueous solution at 25 °C and it can be seen from these figures that hydrogen and hydroxide ions have values far above those of any other ion. This is generally considered to be because of the special relationship between solvated protons and hydroxide ions and the structure of water itself.

Perhaps a more realistic relationship is illustrated in figure 5.6 where the graph of conductivity against concentration is shown for several electrolytes at several temperatures within the range we are considering.

There are some considerations other than conductivity which dictate choice of electrolyte. For example, among the acids commonly available hydrochloric acid can probably be rejected since it may easily lose hydrogen chloride gas when other gases bubble through it or come into contact with it in the electrode pores. Nitric acid has too great an oxidising power and consequently may be too damaging to the cell structure or the electrodes.

Table 5.1 Limiting molar conductivity of some ions in aqueous solution at 25 °C

Ion	$\Lambda°/S\ cm^2\ mol^{-1}$
H^+	349.85
Ca^{2+}	119.0
Ni^{2+}	108.0
K^+	73.50
Na^+	50.15
$(CH_3)_4N^+$	44.92
Li^+	38.64
OH^-	197.6
SO_4^{2-}	160.0
Br^-	78.17
Cl^-	76.35
NO_3^-	71.44
HCO_3^-	44.5
CH_3COO^-	40.9

These figures are obtained by extrapolation of a plot of molar conductivity, Λ, against $\sqrt{c}$ (concentration c) to zero concentration.

Other acids are either too little dissociated (for example, acetic) or too expensive (for example, hydriodic) and really only sulphuric acid remains as a suitable acidic electrolyte, although phosphoric acid, despite its comparatively low dissociation constant, has also been used. Even sulphuric acid has some disadvantages: it is highly corrosive, viscous when concentrated and has a high heat of dilution; it has considerable oxidising power and may yield hydrogen sulphide, a serious catalyst poison, on reduction. Moreover, some electrode systems are not suitable for acidic media, as we have already seen.

Thus we may turn our attention to alkaline electrolytes and will be relieved to discover that in general they are less corrosive than acidic electrolytes (although still having some capacity for attacking metallic and other structures) and have high conductivity. Unfortunately they are unsuitable for any fuel cell system using air rather than oxygen, since air contains a variable but small amount of carbon dioxide which will react with hydroxide ion to give the considerably less conducting bicarbonate ion:

$$OH^- + CO_2 \rightarrow HCO_3^-$$

A further effect is that sodium bicarbonate is not very soluble in concentrated sodium hydroxide solution (a fact which can be employed in the preparation of carbon dioxide free sodium hydroxide solutions), and production of solid sodium bicarbonate may therefore block pores in the electrode and either damage them or at any rate reduce their efficiency.

Potassium hydroxide does not suffer from this difficulty, although it may be slightly more expensive. It is quite soluble in water and a 6 mol dm^{-3} KOH solution (about 30 per cent) has frequently been employed.

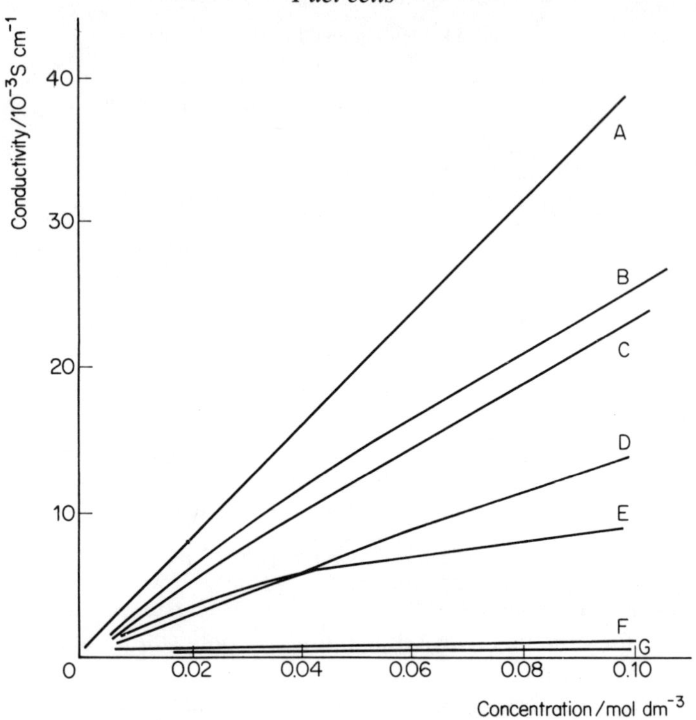

Figure 5.6 Conductivity of aqueous electrolytes as a function of concentration. Key to electrolytes and temperatures:

A	HCl at 25 °C	E	$\frac{1}{3}H_3PO_4$ at 18 °C
B	$\frac{1}{2}H_2SO_4$ at 18 °C	F	CH_3COOH at 18 °C
C	NaOH at 25 °C	G	NH_4OH at 18 °C
D	KCl at 25 °C		

A very concentrated solution of potassium hydroxide in water (about 85 per cent by mass, or 14.3 mol dm^{-3} KOH) has an extremely low vapour pressure (see figure 7.1) and can be used at quite high temperatures since it remains liquid with little or no increase in pressure. A cell using such an electrolyte has been developed for space applications, such as the Apollo missions, as a variant of the Bacon cell described in chapter 7.

Ion exchange membrane electrolytes

In the early nineteen sixties research was carried out with the aim of replacing a free flowing liquid electrolyte by a membrane containing the appropriate ions. Such an arrangement offers the advantages of close contact between the electrolyte and the electrode without any problems of gas escape through the solution. The original intention was to use an ion exchange membrane containing only the ions involved in the electrode processes and no additional ones, but a later modification used added ions of species not concerned in the cell reaction. The ion exchange membrane system offers many worthwhile advantages over the free electrolyte solution type of cell; for example, the membranes have high resistance to gas pressure gradients, low permeability to gases, and a high ability to minimise ion concentration gradients. There are also two interesting technological distinctions: membrane electrolytes can often function well below the normal freezing point of a comparable solution, and they clearly do not suffer from any problems of flow (or lack of it) when used under conditions of zero gravitational field, such as in space vehicles. The solid membrane can also serve as a structural element of the cell and can enable cells to be made which are thin and have the various components packed closely together.

It is important to distinguish between this type of membrane, which actually serves as the electrolyte, and a membrane used as a separator between two electrolyte compartments, an arrangement used in some types of fuel cell.

There are two kinds of membrane structure available: in both types the active part is a cross linked polyelectrolyte containing, for example, sulphonate groups ($—CH_2—SO_2—O^-$, say, which will exchange cations such as H^+), or alkylammonium groups ($—CH_2—N(CH_3)_3^+$, say, which will

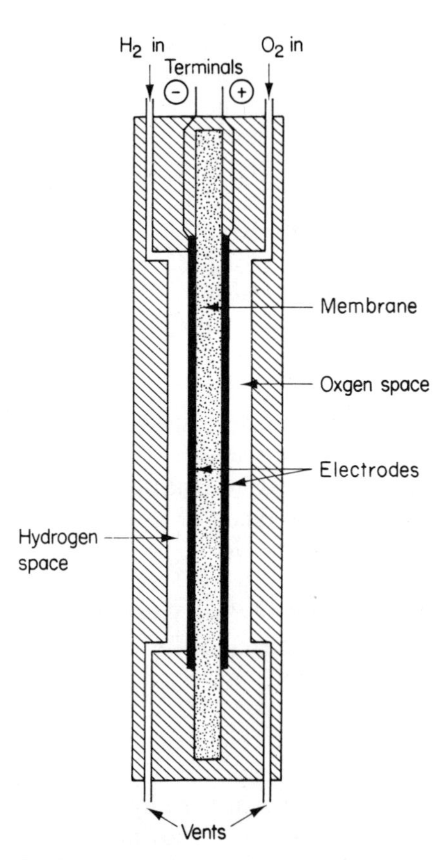

Figure 5.7 Section of an ion exchange membrane cell

exchange anions such as OH^-). However the polyelectrolyte is made in the form of a continuous pore free sheet in the one type, and as beads bonded in some suitable thermoplastic polymer like polyethylene in the other type. In both varieties the conduction is provided by the free ions (cations or anions) held in the membrane by electrostatic attraction to the charged sites on the polyelectrolyte. This conduction is generally less than that for a free flowing solution of the same concentration of a similar electrolyte, but since the path through the membrane is generally shorter than that through any solution used in a fuel cell, this difference may not be important. Moreover, in some applications the membrane is not used as a simple polyelectrolyte with counter ions but is saturated with unbound electrolyte—which may or may not include ions of the species involved in the electrode reactions.

Figure 5.7 shows a diagram of one type of cell containing a membrane electrolyte.

The chemical composition of membranes found to be suitable may be of interest. One example is that of a sheet formed by compression moulding of a partially sulphonated polystyrene at its glass transition temperature (about 160 °C). One of the other type comprises a copolymer of sulphonic acids of polystyrene and divinylbenzene in an inert matrix, probably of poly(vinylidene fluoride). Some improvement in performance has been obtained by use of fluorinated polystyrenes.

There are certain chemical problems, some of which are not yet solved, connected with membrane behaviour. For example, the ionised groups on the polymer may be hydrolysed off in certain circumstances, and the transport of water through or along the membrane depends on a combination of physical and chemical effects.

Capillary cells

Another kind of cell with an immobilised electrolyte is the capillary membrane cell; this has been developed for space use and incorporates what is called static moisture removal. The concentrated alkaline electrolyte of a normal low temperature hydrogen–oxygen cell is confined in a capillary membrane, usually made of asbestos. The pores are of such a size that a very high pressure would be necessary to force liquid out of them, which means that the necessary three phase contact between electrode, gases and electrolyte can readily be maintained even in conditions of low or zero gravitational field and irrespective of orientation. The corresponding pressure for the electrode pores is very much smaller and thus the risk of 'flooding' is low.

The electrodes are made of sintered nickel and are pressed onto the electrolyte membrane. On each side of this sandwich are channels for hydrogen and oxygen gases. Beyond the hydrogen cavity is another electrolyte impregnated asbestos membrane, having a slightly higher concentration of KOH than the main one, and beyond that is a further cavity from which water vapour may be drawn. As the cell operates, water is produced at the hydrogen electrode and the water vapour pressure in the hydrogen supply channels increases. Consequently, water is absorbed into the secondary membrane and can be withdrawn from the far side of this by appropriate adjustment of the applied pressure there. The water vapour can be exhausted, if necessary, straight into space, and of course much waste heat can be dissipated in the same process. A fairly sophisticated control system is, however, required.

Production of water

We have mentioned several times that the chemical product of operation of the hydrogen–oxygen fuel cell is water. There are various problems associated with the production of water which may affect the operation and construction of the cell. In a cell containing a simple aqueous electrolyte, the water formed will clearly dilute the electrolyte unless it can be removed by some means, and if the electrolyte is diluted then, as we have seen, its conductivity will be reduced and hence the current density obtainable must drop.

Surprisingly, this effect is not as serious as it might appear. Experiments have been carried out with hydrogen–oxygen fuel cells containing aqueous alkaline electrolyte over extended periods of time, and although the volume of electrolyte solution has doubled, the power output has not appreciably decreased.

Nevertheless in some circumstances it is important to consider means of removing the excess water produced. We are considering only those cells working in the range between about 20 °C and 100 °C, so clearly working nearer to the higher temperature has the advantage of the relatively easy removal of water by evaporation. The maintenance of these temperatures is usually achieved without transfer of energy from external sources. Cell heating is often an inescapable result of the heat produced by the cell reaction itself, that is that part of the change in enthalpy which is not converted into electrical energy ($T\Delta S$ under equilibrium conditions), and the heat produced by transfer of ions across the bulk electrolyte (sometimes referred to as I^2R heat, from the formula connecting the heat formed by passage of a current I through a resistance R). Thus the problem may well be that of cooling rather than heating.

Hence the water produced may evaporate quite naturally provided that some escape route is provided for it. The escape route is most usually surplus gas leaving the electrodes, which can carry off considerable amounts of water by virtue of its vapour pressure. Thus, providing more gas than can be dealt with by the electrodes may well prove an advantage. The escaping gases can then be cooled, in order to condense out the water, and then recirculated. An alternative arrangement is to remove the water by circulation of the electrolyte itself through an evaporator, or simply to replace it from time to time by a freshly prepared solution. The electrolyte circulation method also provides a means of waste heat removal, should this be necessary—as it often is.

The removal of water from a cell with an ion exchange membrane electrolyte is usually rather easier. If the membrane contains no added unbound electrolyte then the water cannot remove any ions and simply runs off; it can be collected (waste heat can be removed at the same time) by a simple system of absorbent wicks, or it is vented with any waste gases. If oxygen is replaced by air at the oxidiser electrode then the high proportion of nitrogen to be vented naturally carries off the excess water, although there is a danger here that too much evaporation of water will take place and the membrane will dry out and become useless. It is worth mentioning here that ion exchange membranes are not usually stable much above 60 °C, so efficient heat removal is often important. Cells containing a membrane with added electrolyte must be treated somewhat differently: the water produced tends to leach out these unbound ions and therefore the water cannot be drained off but must be evaporated away.

CHAPTER 6

LOW TEMPERATURE CELLS OF OTHER TYPES

We can conveniently make two main classifications of cells working at low temperatures other than the hydrogen–oxygen variety considered in the last chapter, although there may in fact be some overlap. Each type of cell involves the oxidation of materials containing elements other than hydrogen: one of fuels soluble in the electrolyte and the other of hydrocarbons fed to the cell as gases.

Electrolyte soluble fuels

There are several clear practical advantages of cells using liquid fuels dissolved in the electrolyte, over the hydrogen based systems. Electrodes with complicated pore structures which enable contact to be made between catalyst, solution and gaseous fuel can be replaced by much simpler devices in contact only with the solution, although of course a conventional oxygen or air electrode may still be required. Liquid fuels are generally more easily stored than gases; no bulky and heavy cylinders are needed and the energy: mass and energy: volume ratios for the complete power units are likely to be better.

The list of requirements for a fuel of this sort is formidable: it should have freezing and boiling points such that it remains liquid over a wide range of temperatures; it should be soluble in strongly acid and strongly basic media without irreversible chemical reaction taking place; it must be reasonably cheap and easily available; it should be safe to handle; it should react rapidly and without side effects at a suitable positive electrode to give inoffensive products, preferably gaseous, and this reaction should have a high Gibbs function change and a low overpotential; finally it should not react appreciably at the oxidant or negative electrode. Clearly it is improbable that any one substance could prove such a paragon.

Early investigations suggested that organic compounds generally underwent electrochemical reaction more favourably when dissolved in alkaline electrolytes than in acid solution. Study of electrochemical oxidation showed that the most satisfactory substances from the point of view of reactivity were methanol, formaldehyde, ethanol, ethylene glycol and formic acid. Higher alcohols and acids were very much less reactive. Among those substances listed, methanol seems to have the fewest disadvantages, all of the others having severe drawbacks. For example, formaldehyde has a low boiling point (– 21 °C) and is rather unstable; ethanol oxidises to acetic acid which is not particularly volatile; and formic acid reacts in a much less satisfactory way in alkaline solution (when it is, of course, present largely as the formate anion).

Some inorganic fuels have also been examined. Four materials seem to have some promise: ammonia, hydrazine, hydroxylamine and sodium borohydride. However oxidation of the borohydride ion

$$BH_4^- + 10OH^- \rightarrow BO_3^{3-} + 7H_2O + 8e^-$$

which would have a value for $E^\ominus$ of about – 1.2 V, does not proceed satisfactorily because of the hydrogen evolution reaction

$$BH_4^- + 3H_2O \rightarrow BO_3^{3-} + 4H_2 + 2H^+$$

which competes strongly. Moreover, it is expensive as are hydroxylamine and hydrazine. Hydrazine, although explosive and poisonous, has been used in some trial systems with success, and ammonia has also been used.

The methanol fuel cell

The thermodynamic data associated with some of the reactions concerned in the electrochemical oxidation of methanol have been set out in table 2.1. However, computation of the likely emf for a methanol fuel cell has not been wholly satisfactory, since the range of oxidation reactions is quite large. Methanol may be oxidised completely to water and carbon dioxide, although this may be converted by ionisation to carbonate or bicarbonate ions in alkaline solution. It may also be partially oxidised to formaldehyde, which can be hydrated and subsequently ionised:

$$HCHO + H_2O \rightarrow HCH(OH)_2 \rightarrow H^+ + HCH(OH)O^-$$

or to formic acid, which also can be ionised. It will be the precise proportions of these various products, as well as the initial concentration of methanol, which will determine the value of the emf of any particular cell.

Establishment of the potential difference–current relationship for the working cell has not been very satisfactory either, and consequently we do not know much about the mechanism of the oxidation process. All we can say is that the observed potential drop is almost entirely due to activation overpotential, above concentrations where diffusion control is operative. Figure 6.1 shows some experimental results for the limiting current of a methanol cell at two temperatures with acid and alkaline electrolyte.

The important choices for a methanol fuel cell appear to be selection of electrode material and the decision between acid and basic electrolytes. In general the fuel electrodes used have been similar to the metal based

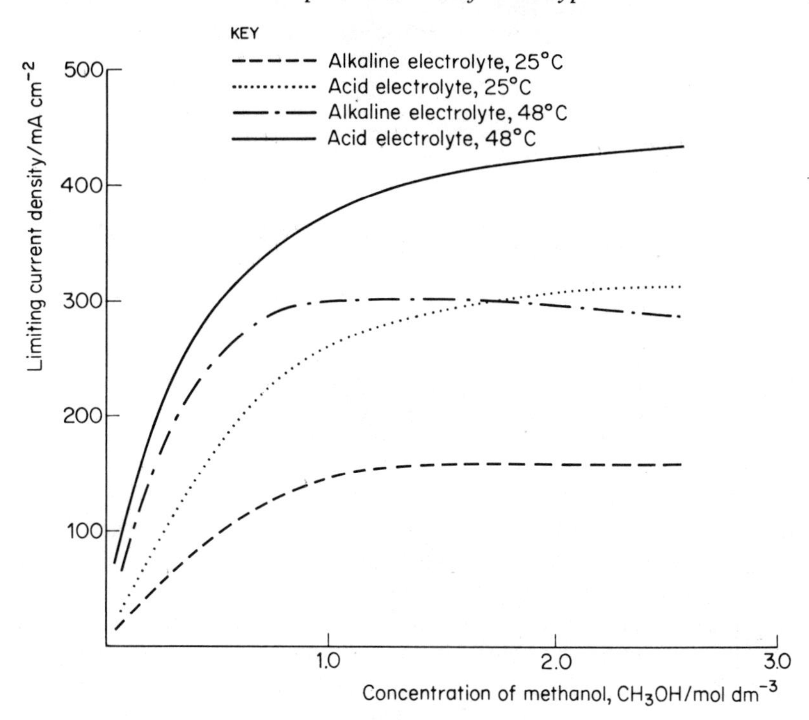

Figure 6.1 Performance of a methanol fuel cell

type described in the previous chapter for hydrogen cells, though often of less complex construction for reasons already mentioned. The base material is nickel or Raney nickel with catalytic amounts of platinum, or a similar metal, deposited on it. The oxygen or air electrode is either silvered metal or an impregnated carbon structure.

The choice between acidic or basic electrolyte is more difficult. It has already been mentioned that the electrochemical oxidation of methanol proceeds more satisfactorily in basic solution than in acidic, but unfortunately we also know that the major product, carbon dioxide, (and even some of the intermediate products, for example, formic acid) react with hydroxide ions to produce a less conducting solution. Moreover this is not the only result of bicarbonate or carbonate ion production; the concentration polarisation will increase because the region nearer the electrode will be progressively more and more denuded of hydroxide ions. Thus with alkaline electrolytes it would probably be necessary to arrange for forced circulation of electrolyte and some kind of external regeneration device. Unfortunately these peripheral pieces of apparatus may require rather a high proportion of the power output by the cell.

Acid electrolytes do not present this kind of problem; any carbon dioxide formed will not be retained in solution but can be immediately vented,

Table 6.1 Operating characteristics of a methanol–oxygen fuel battery

	(a)	(b)
Power output/W	505	750
Total potential difference/V	15	10
Potential difference per cell/V	0.4	0.25
Total current/A	34	70
Current density/mA cm^{-2}	71	150
Power: volume ratio/W m^{-3}	14.6	21.4
Power: mass ratio/W kg^{-1}	7.42	11.0

The figures in column (a) refer to operation in the middle of the range and those in column (b) refer to operation in conditions of maximum overload.
The temperature of operation is 50 °C.

provided a suitable arrangement is made. However methanol does not react so well at the electrode when an acid medium is present, so the argument is certainly not completely one sided. There will of course be water removal problems just as for the hydrogen–oxygen cell.

A further variant worthy of mention is a methanol fuel cell with the oxygen electrode replaced by one containing an oxidant dissolved in the electrolyte. Two kinds have been described. One has electrode compartments separated by a dialysis membrane, and the negative electrode (constructed of Raney nickel–silver) is surrounded by potassium chlorate dissolved in aqueous potassium hydroxide. This cell produces 100 mA cm^{-2} at 0.7 V when operating at 55 °C. Another cell has a particularly neat arrangement of nickel sheets platinised on one side and silvered on the other (comprising both electrodes joined back to back) with an electrolyte of aqueous potassium hydroxide containing methanol and hydrogen peroxide, circulating on each side. It was reported that a forty cell battery of this type produced 40 A at 15 V. The cross section of the electrodes was 645 cm^2.

Finally it may be of interest to look at the operating characteristics of an 'ordinary' methanol cell with platinum–palladium on fuel electrodes, a porous silver–nickel oxygen electrode, and alkaline electrolyte. Some figures are shown in table 6.1.

The hydrazine fuel cell

Hydrazine boils at 115 °C and is toxic and easily detonated. Its hydrate is, however, readily soluble in water and it is not thought to be any more hazardous than petrol for handling and transportation.

Hydrazine is more easily oxidised electrochemically in basic than in acidic solution and the electrode process is presumably

$$N_2H_4 \rightarrow N_2 + 4H^+ + 4e^-$$

or alternatively

$$N_2H_4 + 4OH^- \rightarrow N_2 + 4H_2O + 4e^-$$

According to the information given in table 3.2 the emf of the complete cell (including the oxygen electrode) should be about 1.6 V, that is about 0.3 V greater than the simple hydrogen–oxygen cell. In fact, such a difference is not generally observed, the potential difference obtained from a hydrazine cell being usually about the same as the corresponding hydrogen cell. This suggests that in fact the electrochemical processes are the same and the hydrazine is only acting as a source of hydrogen, possibly by means of initial decomposition on the catalytic surface:

$$N_2H_4 \rightarrow N_2 + 2H_2$$

Some ammonia may also be formed, and this is believed to poison electrode surfaces.

A cell containing a simple palladium activated porous nickel fuel electrode and a 25 per cent aqueous potassium hydroxide electrolyte containing 3 per cent hydrazine, can give about 500 mA cm^{-2} at 70 °C. The electrolyte is recirculated continuously and hydrazine hydrate is added to it during this process to maintain a more or less constant concentration. This circulation also helps the removal of the nitrogen formed as the cell operates. This effluent nitrogen carries off some hydrazine vapour and in view of its toxicity the waste gases are passed through or over acetaldehyde or sulphuric acid to remove any hydrazine. Much of the product water is also carried off with the nitrogen.

Air can be used instead of oxygen but this of course increases vastly the nitrogen flow, and is reported to reduce the power output by about 50 per cent. In all the cells constructed so far there is considerable peripheral equipment which not only requires a fair proportion of the power production but also must be started up before the cell itself is working properly. Thus it seems unlikely that hydrazine cells will ever be important commercially, despite the fact that the energy produced in theory by oxidation of hydrazine is considerably greater than that of most other fuels.

Ammonia fuel cells

Ammonia has a superficial attraction as a fuel since its proportion of available hydrogen is very great if the oxidation reaction is considered to be

$$4NH_3 + 3O_2 \rightarrow 2N_2 + 6H_2O$$

However, the electrochemical reaction

$$2NH_3 + 6OH^- \rightarrow N_2 + 3H_2O + 6e^-$$

has a high overpotential on all catalyst electrodes that have been investigated. One reason adduced for this great contrast with the behaviour of hydrazine is that the nitrogen atoms produced by the ammonia reaction form some sort of surface nitride which tends to poison the catalyst, since the nitrogen atoms cannot readily combine with each other and produce nitrogen gas. The hydrazine molecule already contains two nitrogen atoms linked together and thus no nitride formation is likely. Experimental ammonia cells have been made but they have been found to have a poor performance.

Hydrocarbon fuel cells

It is clear from earlier discussions that any fuel cell able to use hydrocarbon fuels directly without reforming to hydrogen or conversion to methanol or hydrazine will be likely to have a great advantage in cost of materials over the types so far described. Unfortunately, as we know, the direct electrochemical oxidation of hydrocarbons is difficult, although it becomes easier at high temperatures. Cells operating at higher temperatures are described in chapters 7 and 8, but the obvious advantage in a lower operating temperature has led to some investigation of cells capable of working at temperatures of the order of 100 °C, or lower. The kind of fuels that could be used are either gaseous or readily volatile hydrocarbons which can be obtained with minimum impurity from natural gas or oil refinery products. This really restricts the choice to methane, propane, butane and simple olefins.

Table 6.2 shows the electrochemical activity of some hydrocarbons and their derivatives (and also hydrogen) in basic electrolytes. It can readily be seen that hydrogen is far superior to any other fuel and also that the hydrocarbon fuels are themselves considerably inferior to such derived materials as formaldehyde and methanol, which we have already discussed. Nevertheless, some observations are worth making.

In general, the usual considerations apply for choice between acid and alkaline electrolytes. The oxygen electrode behaves much less satisfactorily

Table 6.2 Electrochemical activity of some hydrocarbons and related compounds in aqueous alkaline electrolytes

	Current density/A cm^{-2}	*Potential/V
Hydrogen	0.108	0.7
Formaldehyde	0.054	0.6
Ethanol	0.038	0.5
Methanol	0.032	0.4
Propanol	0.017	0.4
Propylene	0.016	0.5
Butane	0.005	0.3
Ethane	0.003	0.3
Methane	0.002	0.2

*relative to the 'theoretical' oxygen electrode.

in acid solutions but, on the other hand, carbon dioxide rejection is no problem. The alkaline electrolytes will absorb the carbon dioxide formed and regeneration will probably have to be considered. An alternative possibility is to use an aqueous carbonate solution as electrolyte; this will be sufficiently alkaline to allow the oxygen electrode to operate well but will reject carbon dioxide. Unfortunately its conduction and ionic diffusion characteristics will not be as satisfactory as either the acid or the hydroxide solutions. However aqueous solutions of caesium or rubidium carbonate and bicarbonate have been found quite satisfactory, particularly since their high boiling points enable operation to be extended up to 200 °C with consequent improvements in conduction, diffusion and carbon dioxide rejection.

Hydrocarbons are relatively insoluble in electrolytes and so they must either be fed to the fuel electrode as a gas or allowed to form a fourth (liquid) phase. The second possibility is undesirable and could be avoided by adding a suitable detergent material to the mixture. It may also be necessary to prevent the fuel being directly oxidised at the oxygen electrode by inserting a semipermeable membrane, unless the catalytic material chosen for the oxygen electrode is sufficiently selective to prevent any fuel oxidation taking place there.

The fuel electrode is likely to be of the diffusion type as described for the hydrogen cells, and obviously selection of catalyst is most important. If the hydrocarbon is dispersed or dissolved in the electrolyte then the methanol type of electrode can be used.

Low temperature carbon monoxide cells

Carbon monoxide is a product of the steam reforming of hydrocarbons. The simplest case might be represented thus:

$$CH_4 + H_2O \rightarrow CO + 3H_2$$

It was one of a number of fuels tested by the pioneer, W. R. Grove, over 100 years ago. The standard emf values for the oxidation in both acid and alkaline solution are high:

in acid solution: $2CO + O_2 \rightarrow 2CO_2 \quad E^\ominus = 1.33\ V$
in alkaline solution: $2CO + 4OH^- + O_2 \rightarrow 2CO_3^{2-} + 2H_2O \quad E^\ominus = 1.62\ V$

but the actual potential differences obtained for working cells were nowhere near as large as these figures. This has been accounted for by postulating a mechanism which has the conversion of hydrogen to hydrogen ions as the rate determining step at the fuel electrode.

For example, in alkaline solution the following stages are proposed:

$$CO + OH^- \rightarrow HCOO^-$$
$$HCOO^- + OH^- \rightarrow CO_3^{2-} + 2H\ (\text{adsorbed})$$
$$2H\ (\text{adsorbed}) \rightarrow 2H^+ + 2e^-$$

and a similar scheme seems to apply in acid solution, where the initial process is the reaction of carbon monoxide and water. A hydroxide solution would become depleted by replacement of hydroxide ions by carbonate or bicarbonate after operation for some time, and for the reasons mentioned in the previous section attention has been turned to the cell containing aqueous metal carbonate as electrolyte. The electrodes would be of the familiar porous type used with gaseous fuels. Development of this kind of cell seems quite favourable although initial results have not been very encouraging, only low output being obtained. The cell with an acid electrolyte seems less satisfactory, excess carbon monoxide appearing to 'poison' the electrode and thus preventing any high currents being obtained.

Summary of processes occurring in low temperature cells

It is useful to conclude this chapter by listing the processes that must take place when a gas fed fuel electrode operates. Such a list chiefly applies to the hydrocarbon and carbon monoxide cells we have just been examining but, with modification, is relevant both to the dissolved fuel type and also to the hydrogen electrode itself. The stages are:

(1) Diffusion of fuel through dry pores against a flow of product carbon dioxide.
(2) Dissolution of fuel in the electrolyte.
(3) Diffusion of fuel through liquid to a catalytic site.
(4) Reaction at the catalytic site to water and carbon dioxide.
(5) Diffusion of carbon dioxide out through a dry pore against incoming fuel.
(6) Transfer of water to bulk electrolyte.

The process of diffusion of the product hydrogen ions is ignored here.

This summary pinpoints the chief problem over cells using carbonaceous fuels which is the venting of product carbon dioxide by diffusion through the electrode. This may well be the limiting process.

CHAPTER 7

MEDIUM TEMPERATURE CELLS

In this chapter we shall discuss those types of fuel cell operating at intermediate temperatures, that is to say, temperatures higher than those using aqueous electrolytes at atmospheric pressures but lower than those using solid electrolytes. The range is generally considered to be about 150–300 °C. Some kinds of cell operating in this temperature range have already been mentioned during consideration of aqueous electrolyte cells in chapters 5 and 6. In these particular cells, higher temperatures are possible with pressures only slightly greater than atmospheric because of the considerable boiling point elevation obtained with certain highly concentrated solutions, such as the caesium–rubidium carbonate–bicarbonate mixtures mentioned previously. This variety of cell will not be discussed further.

We can, therefore, regard medium temperature cells as those using aqueous electrolytes under considerably increased pressure so that they are maintained as liquids.

It is desirable, of course, to state clearly why there should be interest in operation at higher temperatures since such operation is obviously going to involve complications of design and construction. The reason is simple: chemical and physical processes almost invariably increase in speed with increase in temperature, and the successful operation of high power fuel cells depends on the efficiency of processes such as surface reaction, diffusion of gases and ions and conduction by ions, all of which fall into this category.

Advantages and disadvantages of high pressure cells

The need for increased pressure to prevent loss of solvent from alkaline electrolytes when operating at temperatures in the range 150–300 °C, is clearly shown by figure 7.1.

There is considered to be little advantage in using a solution of potassium hydroxide more concentrated than 45 per cent. This is because it may well solidify at the lower temperatures encountered during periods when the cell is not operating, and thus the system will be effectively like those using molten salts (strictly, the solution will deposit solid KOH on cooling rather than solidifying). Moreover, as figure 7.2 shows, the conductivity of the 45 per cent solution is satisfactory at the temperatures likely to be employed.

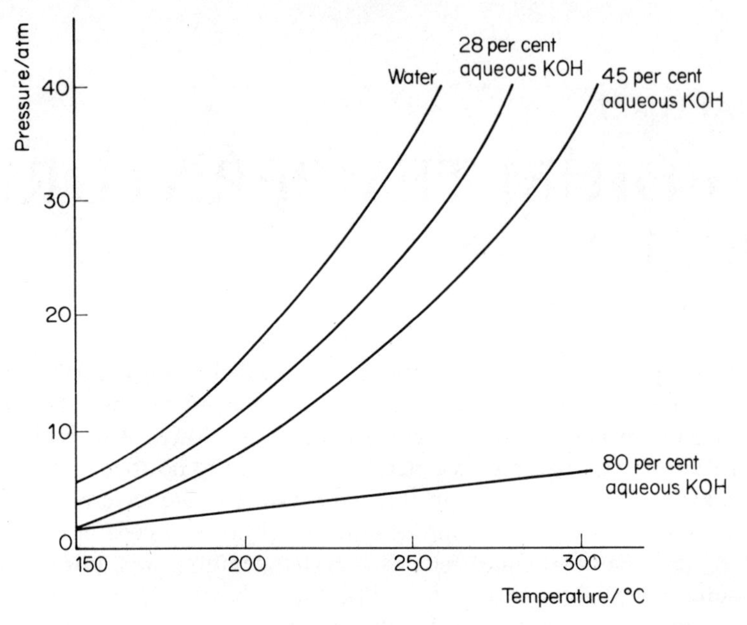

Figure 7.1 Variation of vapour pressure with temperature for aqueous potassium hydroxide solutions

Acid electrolytes have usually not been considered seriously for these cells because the increased temperatures are likely to cause severe corrosion effects.

Some experiments have been carried out with the oxidation of hydrocarbons as the cell reaction, but it has been generally concluded that the simple hydrogen–oxygen reaction is by far the most satisfactory for this type of cell, despite the complications of preparing and storing the pure reactants. Both reactants and the product are gaseous under the conditions of pressure and temperature likely to be encountered in the cell. The reaction itself

$$2H_2(g) + O_2(g) \rightarrow 2H_2O(g)$$

is somewhat less favourable at increased temperature, as has been indicated in figure 2.3; the decrease in Gibbs function change for this reaction will, of course, lead to a decrease in the expected emf, for example, from 1.23 V at 20 °C to 1.09 V at 200 °C. However, as the reaction proceeds from left to right there is a decrease in volume, and so the effect of increase in pressure will be to make the process more favourable. For the example quoted in the previous sentence, therefore, an increase in pressure to 42 atm (27 MN m^{-2}) at 200 °C will bring the expected emf back to 1.19 V. It must be realised that some energy will be required to compress the reactants so

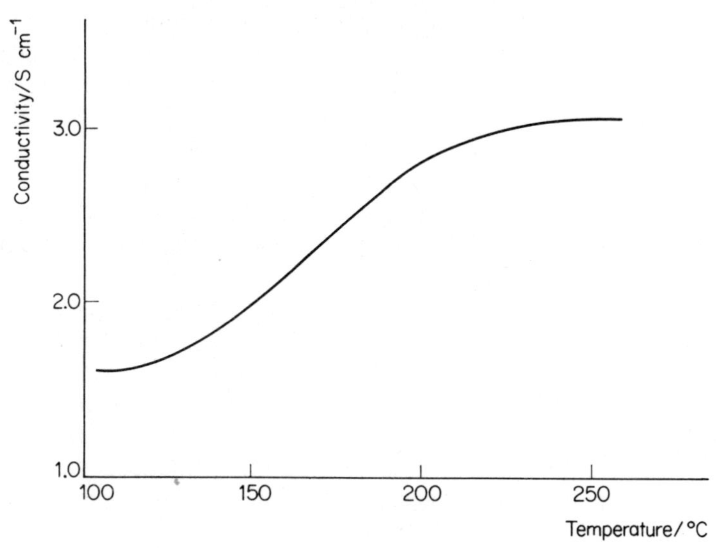

Figure 7.2 Variation of the conductivity of a 45 per cent aqueous potassium hydroxide solution with temperature

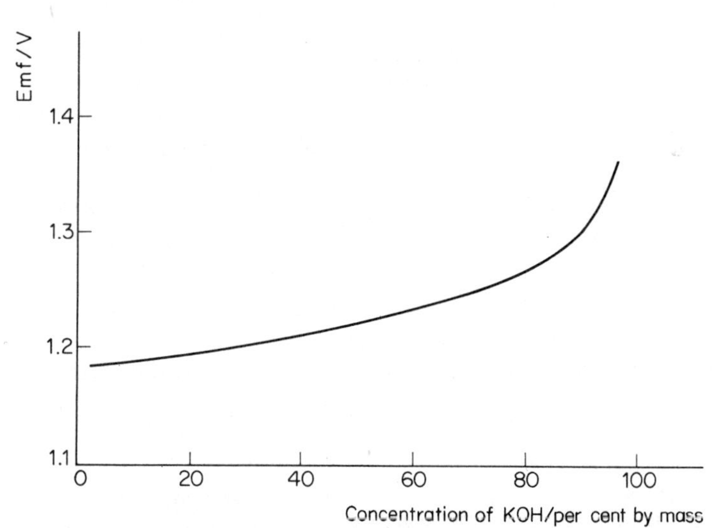

Figure 7.3 Effect of electrolyte concentration on emf for a hydrogen–oxygen cell containing aqueous KOH at 200 °C and 40 atm

any advantages gained by the effect of increased pressure and temperature on the dynamic or polarisation processes must also be significant.

The effect of increasing electrolyte concentration on the emf is entirely favourable, as the graph in figure 7.3 shows clearly. It is not, however, usually feasible to use very high concentrations of potassium hydroxide in water because during cell operation the concentration of water in the immediate vicinity of the oxygen electrode appears to fall to such low levels that the electrolyte becomes solid and impedes proper action of the electrode.

At higher pressures hydrogen and oxygen will be more soluble in the electrolyte which is advantageous although it may lead to more mixing of the reactants with each other and consequently an increased chance of the direct oxidation process taking place. It is also found that under these conditions the material of the oxygen electrode itself tends to be oxidised, unless a particular method of preparation (described later) is employed. Other problems connected with the operation of fuel cells at increased pressures and temperatures are largely mechanical and structural; for example, a fairly complex pressure balancing system has to be adopted for supplying gas to the electrodes. Difficulties of finding sealing materials adequate for making separators and gaskets capable of withstanding hot alkaline solutions in the presence of oxygen under pressure are also quite severe.

Polarisation effects at increased temperatures and pressures

The effect of higher temperatures on all types of polarisation is likely to be favourable. The chemical reactions determining activation overpotential will probably be faster when the temperature increases so that this overpotential will decrease; the diffusion coefficients of both gases and ions will increase, although the decreased solubility of gases may outweigh this to some extent, and therefore concentration overpotential will be lower; and finally the resistance of electrolyte solutions decreases as temperature increases so the ohmic overpotential will decrease too.

The effect of increased pressure is also generally good. Gas diffusion will be increased, although ionic diffusion and ionic conduction will not be appreciably affected. The activation overpotential will probably be reduced, since the higher pressure will lead to a better coverage of the available catalytic sites and hence faster reaction.

A successful working battery of this type of high pressure high temperature hydrogen–oxygen cell was first produced by F. T. Bacon in 1957 after many years of development work. He used a working temperature of 200 °C, an electrolyte of 45 per cent aqueous potassium hydroxide and a pressure of 20–40 atm. A section of such a battery is shown in figure 7.4. The complete battery is of cylindrical 'filter press' shape and contains forty cells or more.

The electrodes of both kinds are initially prepared in the same way from porous nickel, usually as sheets of compressed powder in two sizes of pore, the larger size being on the gas side of the electrode. Thus they are very similar to the type used for the low temperature hydrogen–oxygen system. The fuel electrodes are then treated with an aqueous nickel nitrate solution and heated to produce a layer of oxide which is then reduced to metallic

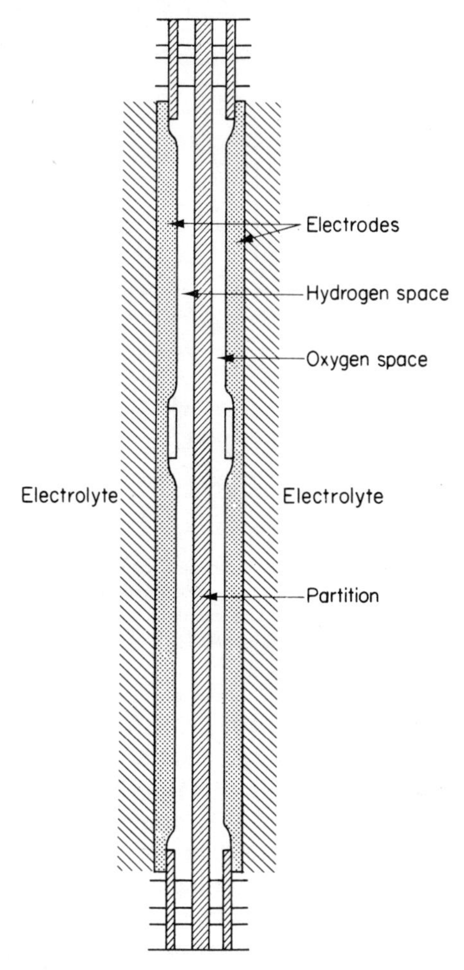

Figure 7.4 Section of one unit of a Bacon high pressure type of cell

nickel by gentle heating in an atmosphere of hydrogen. The oxygen electrode is first oxidised somewhat by heating in air and is then impregnated with an aqueous solution of mixed nickel and lithium nitrates. Final heating then produces the oxides on the electrode surface. Incorporation of lithium in the system reduces the amount of oxidation occurring during the working life of the electrode. Small amounts of noble metal catalysts are sometimes added to the final materials used for oxygen electrode preparation.

Control systems for operation of high pressure cells

A fairly sophisticated control system is needed for this kind of cell in order to balance the various pressures and arrange for the removal of product

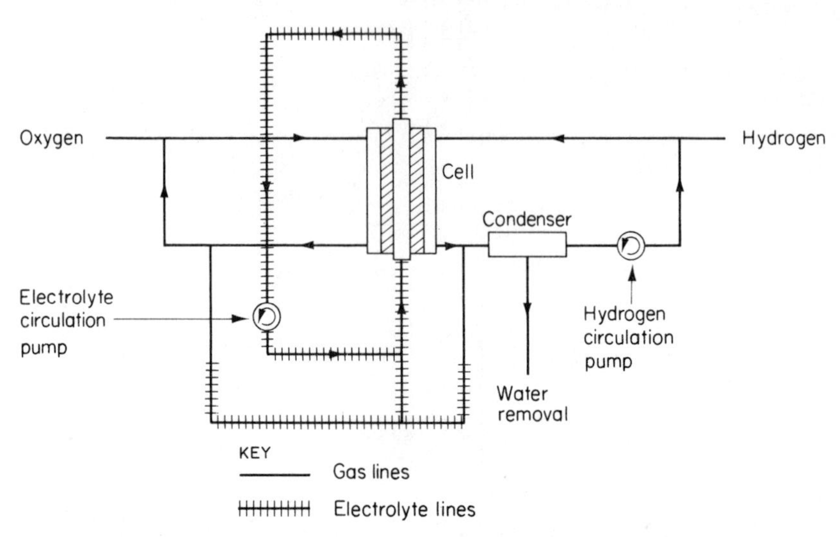

Figure 7.5 Schematic diagram of peripheral equipment required for a high pressure hydrogen–oxygen cell. The pressure balancing arrangement is at the bottom of the diagram. Note the condenser for removing water from the hydrogen circulation system

water and the dispersal of any waste heat produced during operation of the cell. One method of pressure balancing is to allow the hydrogen input pressure to control the oxygen pressure indirectly via the electrolyte circulations system. A schematic drawing of this kind of system is shown in figure 7.5. An alternative method is to control the pressures of the three components by fixing the pressure of an inert gas such as nitrogen which operates suitable balancing valves to equalise the other pressures.

The kind of cell we have been describing is generally well thermally insulated by reason of its style of construction; the reaction occurring within it always produces heat, which has to be removed from time to time. This operation can be effected by circulating hydrogen gas around the fuel electrode in excess of that required for normal working, and this arrangement is usually switched by some form of thermostatic control. As indicated in figure 7.5, the circulating hydrogen passes through a condenser which both absorbs the waste heat and condenses evaporated water. This method is likely to deplete the electrolyte of too much water and some of the condensate must therefore be returned. Again, this requires a control device. Cooling may also be effected by electrolyte circulation either with or without flash evaporation to remove excess water produced by the cell.

Starting up a cell of this sort will almost certainly require preheating and this is usually carried out by resistance electrical heating from an external source, although the battery itself can be switched in sometimes after a certain amount of preheating has taken place.

Performance of high pressure cells

Some typical curves for operation of a high pressure hydrogen–oxygen cell at various temperatures are shown in figure 7.6 and the relation between efficiency and power output and current density is shown in table 7.1. At high current densities not only is the efficiency low but also cell working life is rather short, since the electrodes suffer considerably from corrosion or deterioration due to local heating at reactive sites on the electrodes. Some idea of the relative magnitudes of the different kinds of polarisation losses for this cell are shown in figure 4.7.

Some thought has been given to methods of storing the hydrogen fuel

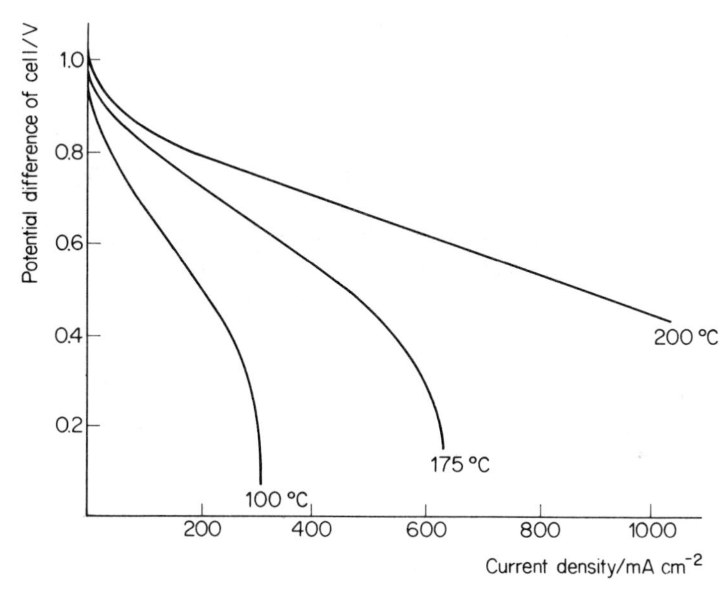

Figure 7.6 Potential difference–current density curves for a high pressure hydrogen–oxygen cell at three temperatures

Table 7.1 Energy efficiency and power output for the Bacon cell

Current density/mA cm^{-2}	26.9	215	645	1614
Potential of one cell on load/V	1.055	0.950	0.820	0.630
Energy efficiency (per cent)	81.5	79.2	68.9	53.0
Power output/kW m^{-2} (per unit area of electrodes)	0.261	2.02	4.28	10.2
Power output/MW m^{-3} (per unit volume of battery)	0.041	0.319	0.830	1.60
Power output/W kg^{-1} (per unit mass of battery)	8.55	66.5	173	333

These figures are for a battery operating at 200 °C and 28.5 atm and with 45 per cent aqueous potassium hydroxide as electrolyte. The 'open circuit' potential of each cell would be 1.188 V.

for this type of cell. Hydrogen would normally be stored in cylinders but these are very heavy for the amount of hydrogen they contain; storage of hydrogen as liquid might be better were it not for the large energy consumption in the liquefaction process. An interesting idea which has been considered is to store the hydrogen chemically by a reversible hydrogenation–dehydrogenation of some suitable aromatic system. For example, cyclohexane decomposes to benzene and hydrogen if heated at about 300 °C over nickel or platinum. The benzene can be reconverted to cyclohexane over nickel at about 200 °C

$$\text{cyclohexane} \underset{200\,^{\circ}\text{C}}{\overset{300\,^{\circ}\text{C}}{\rightleftharpoons}} \text{benzene} + 3H_2$$

cyclohexane benzene

It may be useful to have a circulation system with perhaps a palladium diffusion electrode to separate the hydrogen from benzene and undecomposed cyclohexane.

It is possible, although not at all satisfactory, to operate the cell on fuels other than hydrogen; methane, methanol and carbon monoxide have been tried. 'Town gas' (which is largely a mixture of hydrogen, methane and carbon monoxide) gives a considerable increase in polarisation. Moreover

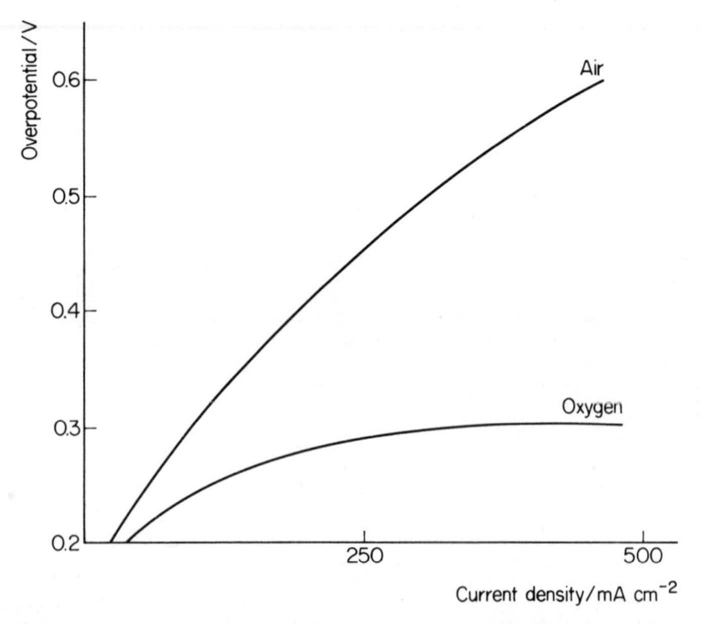

Figure 7.7 Curves for overpotential against current density for air and oxygen supplied to the same electrode. This shows the effect of the nitrogen in the air

carbon monoxide reacts with the electrolyte either before or after oxidation, giving formate or carbonate.

There have also been trials with air in place of oxygen. As might be expected, the high concentration of nitrogen in air causes blocking of the electrode pores and the increase in polarisation is shown in figure 7.7.

It is worth mentioning here that a favoured application of the Bacon type of cell has been considered to be storage of electrical energy by operating the cell in reverse to produce hydrogen and oxygen by electrolysis. This process has been tried and although there appears to be some loss, particularly of oxygen, it could be a satisfactory system. Further discussion on the general topic of fuel cells and electrochemical power storage will be found in chapter 10.

CHAPTER 8

HIGH TEMPERATURE CELLS

We have already seen that in almost all respects the application of increased temperatures to a fuel cell system increases its ease of operation, since almost all physical and chemical processes are speeded up by higher temperatures. In this chapter we shall consider types of cell which have been designed for operation at even higher temperatures than the high pressure aqueous electrolyte variety described in the last chapter. We shall not be surprised to learn that while fuel cell operation becomes easier at higher temperatures, there are vastly increased problems of construction and operation of peripheral equipment such as pumps, heaters and condensers.

There are two main categories of cell designed to run at high temperatures, that is at temperatures of 500 °C or above. One is the cell containing molten salt electrolytes, such as molten carbonates, and the other is the cell utilising a solid electrolyte which may require an even higher temperature of operation.

Molten salt electrolytes

The processes occurring in a hydrogen–oxygen fuel cell operating at higher temperatures without an aqueous electrolyte might well be contemplated as oxide ions produced at the oxygen electrode

$$O_2 + 4e^- \rightarrow 2O^{2-}$$

which then move to the fuel electrode to oxidise the hydrogen

$$H_2 + O^{2-} \rightarrow H_2O + 2e^-$$

and it might therefore be considered that a molten ionic oxide would provide the best electrolyte to encourage this process. However, simple ionic oxides have melting points greater than 1000 °C and therefore attention has been focussed on salts melting at lower temperatures. These salts are generally those with oxygen-containing anions—for example, nitrates, sulphates and carbonates, although some chlorides also melt within the desired range. At high temperatures it is likely that the direct reaction of hydrocarbons at the fuel electrode is quite favourable and hence conversion of petroleum products to the simpler fuels such as hydrogen or methanol

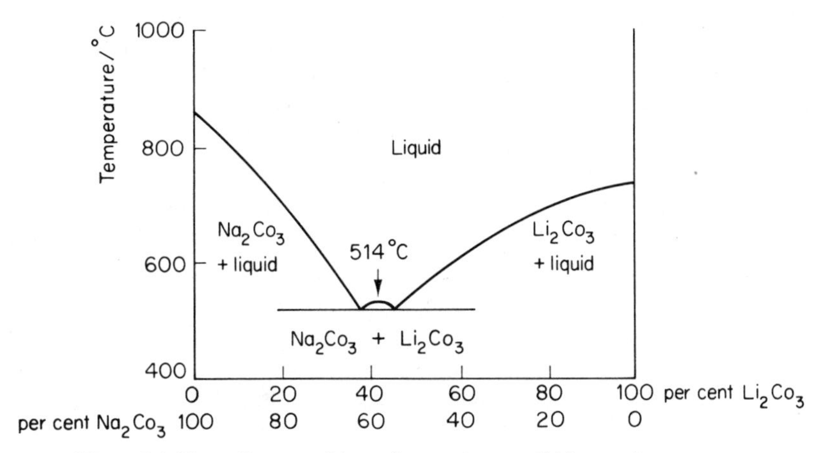

Figure 8.1 Phase diagram of the sodium carbonate–lithium carbonate system

is an unnecessary process and one wasteful of energy. In view of this, consideration must be given to the effect of hydrocarbon oxidation at the fuel electrode on the choice of electrolyte. Carbon dioxide will, of course, be a major product and may well react with any molten salt containing an anion other than carbonate, displacing the appropriate acid anhydride which may then attack the electrode materials or be otherwise troublesome. For example

$$CO_2 + SO_4^{2} \rightarrow CO_3^{2} + SO_3$$

Hence it is most satisfactory to consider as electrolyte a molten carbonate or mixture of carbonates; a mixture of salts may have a considerable advantage since it will have a lower melting point than either component on its own. The phase diagram for the binary mixture of lithium carbonate and sodium carbonate, shown in figure 8.1, well illustrates this point, and it is also of interest to note that a mixture of equal masses of lithium, sodium, and potassium carbonates melts at 390 °C.

A convenient way of maintaining the carbonate composition of the electrolyte invariant is to remove carbon dioxide as a gaseous product from the fuel electrode and transfer it to the oxidant electrode in the air or oxygen stream. Thus for a fuel such as carbon monoxide the overall electrode processes might be

$$O_2 + 2CO_2 + 4e^- \rightarrow 2CO_3^{2-}$$

and

$$CO + CO_3^{2-} \rightarrow 2CO_2 + 2e^-$$

Thus carbonate ion transfer within the electrolyte may be balanced by carbon dioxide transfer outside it. A similar mechanism could operate even for cells using hydrogen as a fuel:

$$H_2 + CO_3^{2-} \rightarrow CO_2 + H_2O + 2e^-$$

Electrodes for molten salt cells

The electrodes to be used in molten carbonate fuel cells at high temperatures are in many ways similar to those for use in other fuel cells. The electrodes must have high catalytic activity and good mass transfer properties, but must also be physically and chemically stable at the temperature of operation and in the presence of fuel, air or oxygen, and the electrolyte. The catalytic and mass transfer characteristics are aided considerably by the high temperatures used but such temperatures of operation also make the possibilities of undesirable chemical reaction rather great, and this is particularly true of corrosive attack by the electrolyte. A further consideration is that the electrodes, like other materials in the system, must be able to withstand the thermal stresses of starting up the cell from the cold and of continued operation at a high temperature.

The form of the electrodes used in experimental cells has been quite varied; powders, gauzes and sintered structures have been employed. A wide variety of metals and metal oxides has been used, the most important consideration being the probability or otherwise of corrosion, particularly at the oxygen or air electrode; above about 750 °C nearly all metals seem to have much the same polarisation effects in this kind of electrolyte, while below that temperature the order of reactivity seems to be

$$Pt > Fe > Ni > Co > Cu > Cr > Mn$$

although this may depend on the particular design of cell and the fuel used.

Flooding of electrodes by electrolyte can still occur and the methods used for low temperature systems (for example, 'waterproofing') are hardly applicable. In order to combat this difficulty two possible devices have been used. One possibility is to 'fix' the position of the electrolyte either by forming it into a stiff paste with magnesium oxide powder or by holding it within a porous magnesium oxide diaphragm. The other possibility is one also used for low and medium temperature cells; that is, to construct the electrodes with large pores on the gas side and smaller pores facing the electrolyte.

Potential and current of molten electrolyte cells

The methods of chapter 3 may be used to establish the emf (or open circuit potential) of any kind of fuel cell. If the simplest fuel, hydrogen, is assumed for a cell with molten carbonate electrolyte, then the cell reaction may be written as

$$2H_2(g) + O_2(g) + 2CO_2(g) \rightarrow 2H_2O(g) + 2CO_2(g)$$

by combination of the individual electrode reactions stated on p. 85. It is necessary to distinguish between the carbon dioxide regarded as 'reactant' (left hand side of the equation), which is that consumed at the cathode, or oxygen electrode, and the carbon dioxide regarded as 'product' (right hand side of the equation), which is that produced at the anode, or hydrogen electrode, since they are most likely to have different partial pressures.

We can show quite simply that the emf of this cell is given by

$$E = E^{\ominus} - \frac{RT}{4F} \ln \frac{P_{H_2}/\text{atm}^{-1}}{P_{H_2}^2 P_{O_2}} - \frac{RT}{2F} \ln \frac{P_{CO_2}(a)}{P_{CO_2}(c)}$$

where $P_{CO_2}(a)$ and $P_{CO_2}(c)$ are the partial pressures of carbon dioxide at anode and cathode (as defined above) respectively and the other symbols used have their conventional meanings.

However, as we know, the actual working potential of the cell will depend on the effects of polarisation which, as usual, are conveniently divided into three. Concentration polarisation arising from gas diffusion effects is not usually very significant but unless carbon dioxide is supplied to the oxygen electrode in the manner described earlier, polarisation—because of ionic movement in the electrolyte—can be quite substantial.

Activation polarisation for cells using hydrogen or carbon monoxide or a mixture of the two operating between 550 °C and 700 °C is essentially zero, but if hydrocarbon fuels are used below about 600 °C then some overpotential is observed and thermal cracking also gives carbon deposition which fouls the electrodes. The most significant contribution to the overpotential of these cells with molten carbonate electrolyte is that arising from ohmic polarisation. The resistance of the electrolyte itself is usually significant and this figure will be increased quite markedly by any supporting structure or diaphragm used.

Design of molten electrolyte cells

Cells of this type were first made in the nineteen twenties, but the earliest design (by the Swiss chemist, Bauer) gave an unsatisfactory power output and had a short life. A better design was produced about 30 years later by Broers, working in the Netherlands, who based his initial work on the solid electrolyte cell of Davytan (discussed in a later section). A sectional diagram of the Ketelaar–Broers cell is shown in figure 8.2. It operates in the temperature range 500 °C–900 °C and the fuels used have been hydrogen, carbon monoxide, methane, and various mixtures of these substances. The electrolyte is a mixture of lithium, sodium and potassium carbonates (in various proportions) supported on porous magnesium oxide discs, prepared by sintering the commercial product at 1200 °C. The electrodes are made of

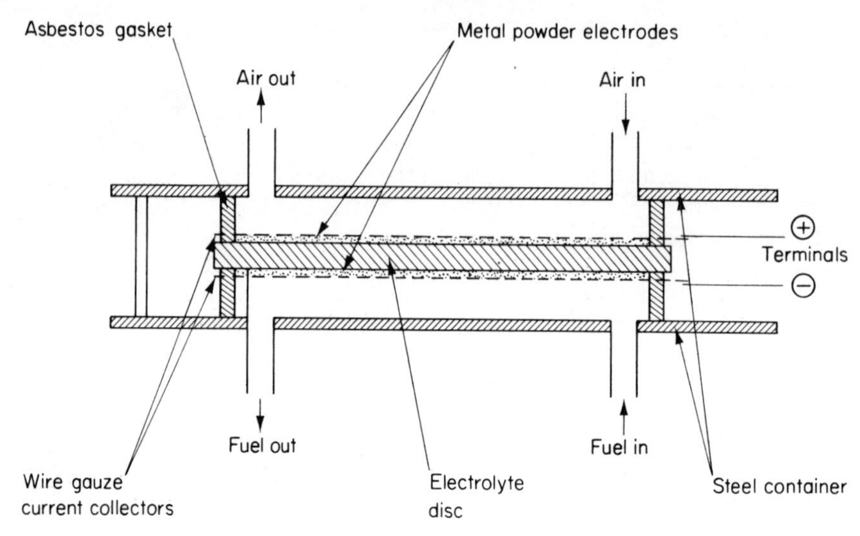

Figure 8.2 Section through the experimental Ketelaar–Broers cell

fine metal powder (silver for the air electrode, and various metals including platinum, nickel, copper, chromium and manganese for the fuel electrode) backed by a gauze (generally of the same metal) and a stainless steel plate. In order to use methane satisfactorily as a fuel, it was necessary to use temperatures above 750 °C and add water vapour to the methane: this suggests that steam reforming takes place before any electrochemical reaction

$$CH_4 + H_2O \rightarrow CO + 3H_2$$

$$CO + H_2O \rightarrow CO_2 + H_2$$

This type of cell had quite good operating characteristics and had a lifetime of several months, unless the temperature rose above about 750 °C when the electrolyte tended to lose carbon dioxide and even metallic oxides by evaporation, and the gaskets were chemically attacked.

Several variations on this design have been developed. In some cases different electrode materials have been used, for example, the Bacon type electrodes with two pore sizes. The electrolyte has been generally similar but in many applications carbon dioxide has been introduced with the air or oxygen for reasons discussed in an earlier section. The best cells have produced power outputs of about 60 mW cm^{-2} for a potential difference of about 0.6 V. Cell life has usually been terminated by cracking of the supporting discs allowing loss of electrolyte or mixing of fuel and air, or by corrosion of gaskets or the containing materials. Some improvement has been obtained by using a paste electrolyte of the molten carbonates mixed with magnesium oxide powder; in some cases this has been combined with a concentric mode of construction, one electrode being a tube down the centre of a cast cylinder of paste electrolyte and the other electrode being a tube outside it.

Molten electrolyte cells operating at high temperatures appeared to offer considerable advantages over the low temperature systems, in that the electrochemical oxidation of fuels would take place very readily without the need for expensive electrode materials. Unfortunately this hope has been vitiated somewhat since the discovery of severe corrosion problems for this sort of medium at the temperatures used.

Solid electrolytes

The disadvantages of using molten salts as electrolytes, as outlined in the last section, can largely be overcome by use of solid electrolytes. It must be made clear immediately that these materials conduct electricity by an ionic mechanism and not by movement of electrons, as a metal or a semiconductor would. A material having electronic conduction would of course 'short circuit' any fuel cell reaction, and is therefore not suitable.

Zirconia, or zirconium (IV) oxide, ZrO_2, when pure is a typical insulator, its resistivity decreasing from about 10^{14} Ω cm at room temperature to

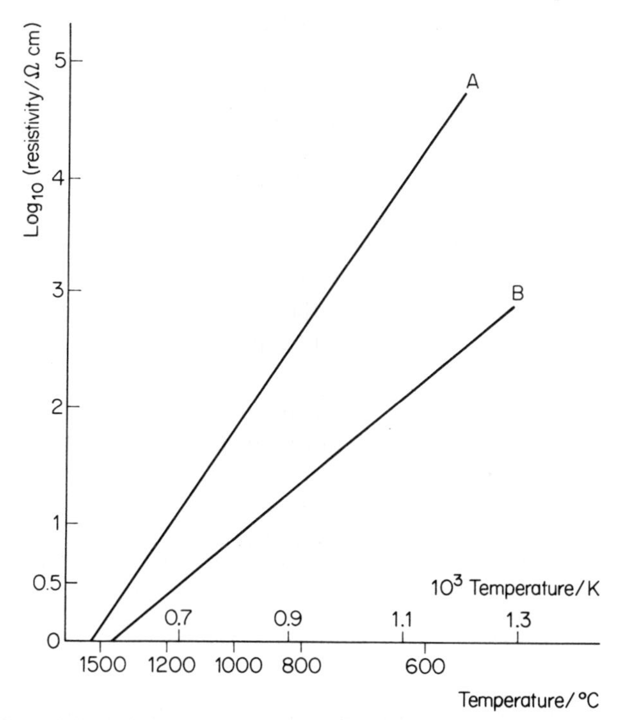

Figure 8.3 Variation of resistivity of two zirconia samples with temperature. A = 85 per cent by mass ZrO_2, 15 per cent by mass CaO; and B = 90 per cent by mass ZrO_2, 10 per cent by mass Y_2O_3

about 10^7 Ω cm at 1000 °C. When small quantities of calcium oxide, CaO, and yttria, yttrium (III) oxide, Y_2O_3, are added to zirconia its resistivity drops to about 50 Ω cm at 1000 °C and decreases rapidly with increasing temperature. The variation of resistivity with temperature for two treated zirconia samples is shown in figure 8.3.

It is supposed that the mechanism of conduction through the zirconia structure is as follows. When calcium oxide or yttria are mixed with the zirconia, some Ca^{2+} ions or Y^{3+} ions displace Zr^{4+} ions from their positions in the lattice. Because of the different charges on the replacing ions, this process leads to a certain number of oxide ion lattice sites becoming empty. At high temperatures it is possible for oxide ions to migrate through the lattice via these vacant sites and hence the structure conducts electricity by an ionic mechanism. Experimental investigations have shown that the conductivity is due almost completely to movement of oxide ions: there is no appreciable movement of positively charged ions, and less than 2 per cent of the conductivity is due to movement of electrons. The observed ionic mobility, L, depends directly on the diffusion coefficient of the oxide ions, D, according to the Nernst–Einstein relation

$$L = \frac{Dez}{kT}$$

where z is the charge number of the ion, e the charge on the electron, k the Boltzmann constant, and T the thermodynamic temperature.

The maximum conductivity is reached for mixtures of about 12–13 mol per cent CaO in ZrO_2 and of about 8–9 mol per cent Y_2O_3 in ZrO_2 at 1000 °C. The resistivity of the second mixture is about one fifth that of the first. The addition of other oxides—for example those of ytterbium, niobium and scandium—has also been examined.

A further point of great significance about these mixed oxides is that the additions seem to stabilise the tetragonal or fluorite form of zirconia, nomally only thermodynamically stable above 1150 °C. This transition of pure zirconia at 1150 °C makes it structurally weak on repeated heating and cooling; in other words it tends to break up on thermal cycling. Apart from this point it would be most satisfactory as a solid electrolyte: it is chemically very stable under both oxidising and reducing conditions, and is impermeable to gases at high temperatures. Consequently, the improvement in conductivity and structural properties caused by the addition of calcium or yttrium oxide, make it almost ideal. So much so in fact that there has been little enthusiasm for searching for any other materials suitable for solid electrolyte fuel cells.

Construction of solid electrolyte cells

The requirements for electrode manufacture are generally the same as for other high temperature cells; the material used must be chemically and physically unaffected by the temperature of operation but must be porous to the fuel or oxidant used. The method of preparation used is generally to coat the electrolyte disc itself with the substance chosen as an electrode. Because of the relatively low conductivity (compared with a liquid electrolyte) of the zirconia discs, it has usually been thought desirable to make these as thin as possible provided that they remain impervious to the gases used. The thickness used may be as little as 0.1–0.4 mm.

Some metals such as silver can be evaporated in a high vacuum and allowed to condense on the electrolyte surface. Alternatively, a dispersion in some suitable organic binder, which is readily decomposed on heating, can be applied to the electrolyte surface so as to leave the necessary porous conducting metallic film. Another method is to operate in reverse and form the electrolyte itself on the metal electrode, by vaporisation of zirconia with a high energy electron beam followed by condensation on the selected sintered metal surface. A quite different idea is to use a porous carbon

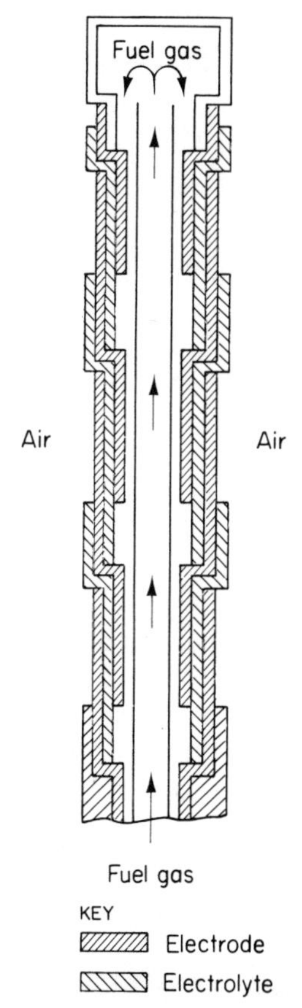

Figure 8.4 Section of a cylindrical high temperature cell

electrode and produce it *in situ* by using natural gas as a fuel and ensuring that electrochemical oxidation is incomplete. Certain metals cannot be used since they tend to become incorporated in the electrolyte and change its mode of conduction to the electronic pattern; nickel is an example of one of these. Nickel oxide is, of course, well known to be semiconducting.

As we have indicated already this type of fuel cell is generally made up of a series of electrolyte discs between electrodes. The outer faces of the electrodes will be in contact with air or oxygen on the one hand and fuel on the other. Another method of construction is to use a cylindrical arrangement with (for example) fuel on the inside of the concentric tubes and air on the outside. The whole device must be placed in a furnace to maintain the necessary working temperature although to some extent this may be self sustaining. A diagram of the cross section of an interesting form of a cylindrical cell is shown in figure 8.4.

Potential and current relations for a solid electrolyte cell

The cell reactions for solid electrolyte cells may be represented as either

$$2H_2 + O_2 \rightarrow 2H_2O$$

or

$$2CO + O_2 \rightarrow 2CO_2$$

or possibly both together. Thus the emf of each type of cell (which may be considered separately) will be given by the expressions

$$E = E^{\ominus} - \frac{RT}{2F} \ln (P_{H_2O}/P_{H_2}) + \frac{RT}{4F} \ln (P_{O_2}/\mathrm{atm})$$

and

$$E = E^{\ominus} - \frac{RT}{2F} \ln (P_{CO_2}/P_{CO}) + \frac{RT}{4F} \ln (P_{O_2}/\mathrm{atm})$$

by the methods of chapter 3 (ideality of gas behaviour being assumed). Calculations carried out for the hydrogen–oxygen system, using the first of these equations, shows good agreement with experimental measurements, as can be seen from figure 8.5.

As we discussed earlier for molten electrolyte cells, activation and concentration polarisation are negligible at these high temperatures of operation. The only significant drop of potential from the 'open circuit' value (or emf) is that due to the resistance of the solid electrolyte (the 'I–R' drop, or ohmic overpotential), which unfortunately can be quite considerable since, as has already been made clear, solid electrolytes do not have quite such desirably high conductivities as the liquid variety. However, as we have seen, this effect can be minimised by using very thin films of electrolyte in working cells. The relation between current and potential difference for this kind of

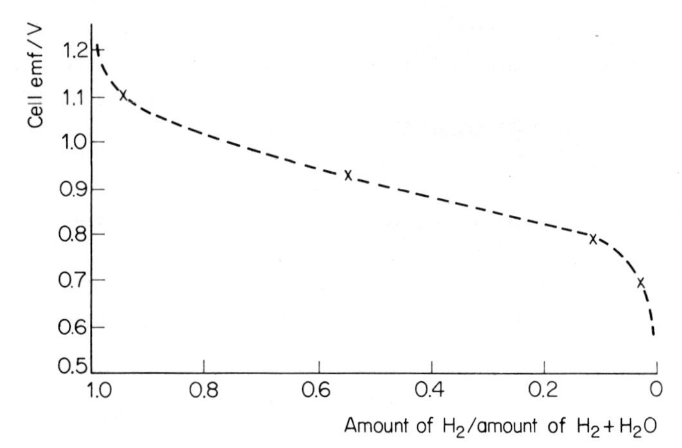

Figure 8.5 Comparison of measured emf of a high temperature cell with that calculated from the equation

$$E = E^{\ominus} - \frac{RT}{2F} \ln \left(\frac{P_{H_2O}}{P_{H_2}} \right) + \frac{RT}{4F} \ln (P_{O_2}/\text{atm})$$

Experimental points are indicated by crosses and the broken line is the calculated curve. The cell has platinum electrodes and an electrolyte of 85 per cent by mass ZrO_2 + 15 per cent by mass CaO. It operated at 1015 °C with an oxygen pressure of 731.2 mm Hg

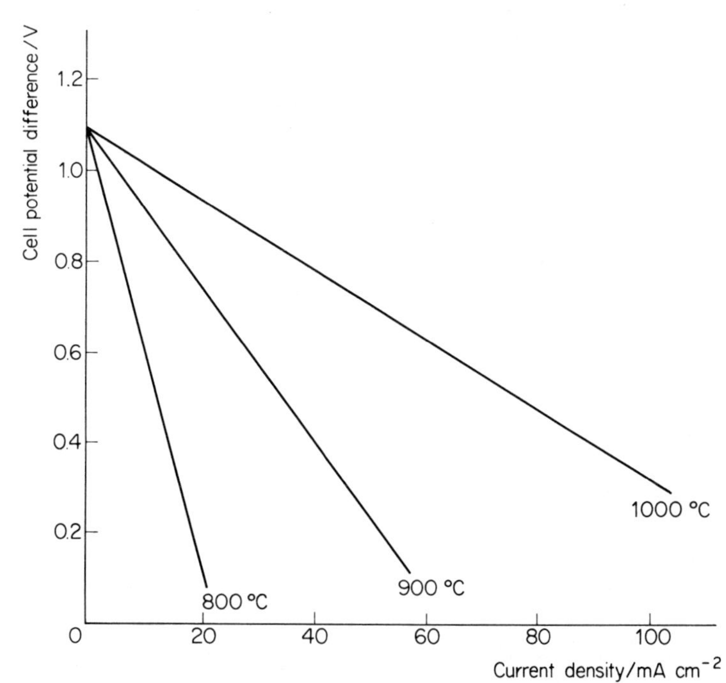

Figure 8.6 Potential difference–current density graph for the hydrogen–oxygen solid electrolyte cell at three temperatures of operation

cell (hydrogen–oxygen with 0.4 mm thickness of 85 per cent ZrO_2 plus 15 per cent CaO electrolyte) at various temperatures can be seen in figure 8.6. The fact that all the lines on this graph are straight is a further indication that only ohmic overpotential is concerned, since Ohm's Law is clearly obeyed (current flowing is proportional to the drop in the potential difference).

This type of cell offers some prospect of importance in the future provided that the problems over choice of materials, both for electrodes and for gaskets, can be solved. The electrolyte system seems quite satisfactory and catalytic problems are minimised by the use of the high temperatures.

CHAPTER 9

AIR DEPOLARISED CELLS AND OTHER CELLS OF INTEREST

Air depolarised cells

Some mention was made in chapter 1 of the air depolarised cell; it is certainly worth considering in a book on fuel cells since, although the 'fuel' used—the metal electrode—is not at all a conventional fuel, there are nevertheless considerable resemblances. It was pointed out in chapter 1 that one of the important characteristics of a fuel cell was that the material used up in the operation of the cell was almost entirely supplied to it and was not built in at the time of construction, whereas other kinds of galvanic cells usually could only operate on the materials actually put into the cell during its manufacture. The air depolarised cell occupies an interesting position in between these extremes; the oxygen or air electrode clearly operates from material supplied during operation, while the fuel electrode is a metal one manufactured as part of the cell and converted to oxide or other products during its life. The simplest and most well-known air depolarised cell is the zinc–air cell with an alkaline electrolyte:

$$Zn(s) \,|\, K^{+}OH^{-}(aq) \,|\, O_2(g) \,|\, C(s)$$

which has a cell reaction

$$2Zn + 4KOH + O_2 \rightarrow 2K_2ZnO_2 + 2H_2O$$

the potassium zincate being of course an oxidation product of the zinc electrode.

This kind of cell was discovered in 1879 by Maiche. He modified a Leclanché cell by replacing the carbon–manganese dioxide positive electrode of that cell with a platinum–carbon one, thus using oxygen or air as a 'depolariser' rather than manganese dioxide. In fact, Leclanché himself, it seems, had already noticed that his ordinary cell worked rather better when air had access to the positive electrode material. A commercial version of this cell was produced in the nineteen thirties and has been quite extensively used in locations where mains electricity supplies were impractical, in particular for railway signalling and radio operation. Subsequently a 'dry' version, with an immobilised electrolyte, has been developed.

Air depolarised cells clearly share with fuel cells the problems of the oxygen electrode discussed in chapter 4, but there are fewer problems over availability of fuel or its preparation, and the production of carbon dioxide and water during operation is either non-existent or less serious.

The air electrode

The kind of electrode employed in these cells is very similar to those described in chapters 5 and 6 where a porous electrode in contact with both gas and electrolyte was required. The material of the electrode is porous nickel or carbon impregnated with silver, or another metal, as a catalyst. Platinum group metals cannot be used in metal–air cells, apparently because of the risk of migration to the metal electrode where a kind of short circuit can be set up. Electrolyte flooding of the electrode pores is generally prevented by a waterproofing treatment, since the electrode with two sizes of pore used extensively in fuel cells (see chapter 5) is not suitable for arrangements where the air or oxygen is not supplied under pressure, and this is almost invariably the case for metal–air cells.

The metal electrode

The choice of the metal used for the other electrode depends on the balancing of various characteristics, such as ease of oxidation of the metal, its stability in contact with the electrolyte and its energy density (that is electrical energy produced per unit mass of metal). Naturally there are the usual considerations of cost and availability. The emf values calculated for various metal–air cells are shown in table 9.1 which also gives the energy density figures.

Zinc is usually considered the most suitable for these cells because it has a high potential but its rate of reaction in absence of electrochemical activity can be reduced with a suitable inhibitor—for example, amalgamation of the surface by mercury. This means corrosion is not appreciable when the cell is not in use.

Table 9.1 Electromotive force and power: mass ratios for some possible metal–air cells

System	Emf, $E^{\ominus}/V$	Power: mass ratio/W h g^{-1}
Lithium–oxygen	3.44	13.26
Aluminium–oxygen	2.70	8.03
Magnesium–oxygen	3.09	6.81
Zinc–oxygen	1.65	1.35
Iron–oxygen	1.28	1.22
Cadmium–oxygen	1.21	0.57
Lead–oxygen	1.59	0.41
(Compare:		
Lead–acid	2.14	0.25)

A further consideration is the nature of the oxidised product. Zinc forms either an oxide or a hydroxide and usually conditions can be arranged so that no solid is formed but the material goes into solution, for example, as the zincate. Formation of solid may block electrode pores and reduce the output of the cell. One of the important concerns is the extent to which the metal electrode retains low polarisation characteristics during the life of the cell. For this reason it may be necessary to incorporate a current carrying grid made of an inert metal to maintain electrode conduction as the metal is electrochemically oxidised.

Metals such as aluminium and magnesium have also been tried, but they tend to be rather too reactive and hence cells constructed with them have a poor shelf life.

Electrolytes for metal–air cells

Considerations are very similar to those for fuel cells. Aqueous acidic or alkaline electrolytes are preferred because of their relatively high conductivities and, of these, the alkaline electrolyte is certainly the better medium for operation of the oxygen electrode. There are the usual difficulties about the absorption of carbon dioxide from the air supplied, which causes depletion of hydroxide ion in the electrolyte, and there is an additional problem connected with evaporation of water through the air electrode.

Metal–air cells have been made in what can be termed 'mechanically rechargeable' forms; when the metal electrode has been essentially used up or the electrolyte saturated with product or otherwise vitiated, the cell can be emptied, the old negative electrode removed and replaced by a new one, and the cell then filled up again with fresh electrolyte. The air electrode is unlikely to decay appreciably and therefore need not be changed. It may be possible to recover metal from the exhausted electrolyte by a chemical reduction process.

Secondary cells which, like the lead–acid battery, can be electrically recharged have also been investigated, although there do seem to be some difficulties. The essential idea is to regenerate the zinc from the dissolved zincate by passing a charging current through the cell. The problems are that the zinc may grow in an unbalanced way, either producing electrodes of a curious and inefficient shape or causing them to form tree-like growths which provide short circuits to the air electrode. The air electrode also behaves badly under charging conditions; the evolutions of oxygen may disrupt the pore structure of the electrode and, more seriously, may oxidise the materials of its constructions. A possible way of avoiding this particular difficulty is to incorporate a third, charging electrode in the cell so that the air electrode remains idle during the process. Unfortunately, this answer to the problem leads to considerable losses in efficiency because of the presence of these additional electrodes.

Sodium amalgam cells

One of the metals not shown in table 9.1 which would have a high emf if it could be used in a metal–air cell, is sodium. However, as everyone with any knowledge of chemistry knows, metallic sodium reacts rapidly and, in some cases, even explosively with water at ordinary temperatures, which presumably means no possible use could be made of this high theoretical emf. However, a dilute sodium amalgam (that is, a solution of sodium in mercury) is apparently stable in contact with aqueous electrolytes, and this has been made the basis of a variety of metal–air cell. The amalgamation of sodium reduces the magnitude of its standard electrode potential considerably (– 2.714 V to – 1.957 V) but the main effect is a large increase in the overpotential for hydrogen evolution at the amalgam surface. However, the actual emf of the cell

$$\mathrm{Na(Hg)\,|\,Na^+, OH^-(aq)\,|\,O_2(g)}$$

depends on the activity of the sodium in the amalgam as well as on the activities of the other species in the cell. The methods of chapter 3 give the relation as

$$E = E^{\ominus} + \frac{RT}{4F}\ln a_{\mathrm{Na}} + \frac{RT}{4F}\ln\frac{a_{\mathrm{O_2}}a^2_{\mathrm{H_2O}}}{a_{\mathrm{Na^+}}}$$

for the cell reaction

$$\mathrm{4Na(Hg) + O_2(g) + 2H_2O(\ell) \rightarrow 4NaOH(aq)}$$

The sodium amalgam is a notably non-ideal solution and therefore the effect on the emf of change in its concentration is very marked. However, a concentration of Na in Hg higher than about 4 mol per cent cannot be used since direct reaction with the electrolyte then takes place.

The cell developed by E. Yeager has a negative electrode composed of a steel plate down which amalgam runs in a continuous film; it is collected in a sort of sump and recycled. This electrode shows practically no activation polarisation and very little concentration polarisation. The oxygen electrode is a hollow block of porous carbon treated with a silver active carbon preparation. It is important to make sure that this electrode does not shed particles of carbon since these seem to have a substantial effect on the amalgam electrode, reducing the overpotential for hydrogen evolution considerably. It is also necessary to prevent air having contact with the amalgam. A single cell of this type gave 200 mA cm^{-2} at 1.42 V with 0.52 per cent by mass sodium amalgam and 5.1 mol dm^{-3} aqueous NaOH at 25 °C.

Although this cell has never achieved commercial viability there are certain attractive features about it, notably its high potential difference, its operation at ordinary pressures and temperatures, the ease of storage of 'fuel', and the lower oxygen requirements. There are, however, also considerable disadvantages: sodium and mercury are expensive; mercury is very poisonous; and the cell requires quite complicated auxiliary equipment for circulation of amalgam and of electrolyte and for the various control systems.

Biochemical fuel cells

The human body has sometimes been likened to a fuel cell, the fuel being food oxidised in the body to various waste products and producing useful energy, some of which is electrical. It is not unreasonable therefore to consider using a biochemical reaction as the basis for an experimental fuel cell. There has been particular interest in this idea because of the large quantities of vegetable matter available and also because of the possibility of using biological waste produced in spacecraft. The second of these ideas is of significance not only because of the biological material being a possible fuel for making electrical energy but also because it may help to conquer the problem of what to do with such waste in a closed environment such as a space vehicle.

It is possible to conceive of two main varieties of biochemical fuel cells. The *indirect* type uses the biological biochemical reaction to convert the biological 'fuel' (for example, excretion products and other waste) into substances which can then take part in a conventional electrochemical process, such as hydrogen, ammonia or oxygen. Apart from the fact that it is advantageous for the biochemical reaction to take place as near the site of the subsequent electrode reaction as possible, these fuels do not differ seriously from the kinds we have discussed in earlier chapters.

The *direct* type of biological fuel cell is one in which the electron transfer reaction takes place to or from the biochemical material directly. This sort of redox system is not normally catalysed by metals but requires biochemical catalysts such as enzymes. Such catalysts are frequently specific and will effect reactions of considerable interest in fuel cell systems. For example, the partial oxidation of glucose to gluconolactone

$$C_6H_{12}O_6 \rightarrow C_6H_{10}O_6 + 2H^+ + 2e^-$$

is catalysed by the enzyme *glucose oxidase*, and the reduction of nitrate ions to nitrogen is effected by the organism *micrococcus denitrificans:*

$$2NO_3^- + 12H^+ + 10e^- \rightarrow N_2 + 6H_2O$$

The organisms which are concerned in some reactions of this sort are living and therefore need some energy, which would detract somewhat from the power output characteristics of any cell using the reaction. Alternatively, a 'dead' preparation may be used, in which case the catalytic medium would have to be renewed from time to time.

The superficial attractiveness of this kind of fuel cell is clear; moreover there seem to be other advantages.

Table 9.2 Standard electrode potentials of some bioelectrochemical reactions

Reaction	Catalyst	$E^{\ominus}$/V
$CO(NH_2)_2 \rightarrow NH_3$	micrococcus ureae	− 0.47
Carbohydrate → ethanol	saccharomyces	− 0.37
$CO_2 \rightarrow CH_4$	methanobacillus	− 0.25
$SO_4^{2-} \rightarrow H_2S$	desulphovibrio desulphuricans	− 0.22
$NO_3^- \rightarrow N_2$	micrococcus denitrificans	0.75
$CO_2 + H_2O \rightarrow O_2$	algae photosynthesis	0.82

These 'standard' potentials are actually referred to pH 7, i.e. neutral aqueous solution, and are at 25 °C.

(1) The fuel, in principle, could be provided by a wide variety of natural products or waste animal and vegetable matter.
(2) The oxidation may be effected by inorganic anions commonly present in, for example, sea water.
(3) No high temperatures or concentrated electrolytes are needed.
(4) In many cases catalysts can be continuously regenerated.

There are of course some disadvantages: table 9.2 gives the estimated standard electrode potentials for some bioelectrochemical reactions. It can be seen that these are all rather low. Moreover such potentials can rarely be achieved, since conditions for successful electrochemical reaction (giving low overpotentials) and those for most biochemical processes are incompatible. The parameters determining the kind of conditions concerned are temperature and the various electrolyte concentrations. In fact we can summarise the disadvantages.

(1) High electrolyte concentrations (which aid conduction) are usually detrimental to biochemical preparations.
(2) Biologically active materials only dissolve in water to a limited extent while retaining their activity.
(3) Biologically active materials (for example, enzymes) are rather easily (and irreversibly) 'poisoned' by traces of other substances.
(4) Although increased temperature may improve rates of reaction it also hastens the 'denaturing' of enzymic materials, in which process they lose their catalytic activity.

Thus the prospects for biochemical cells are not good; their current densities might be expected to be about 1 per cent of the conventional 'inorganic' type of cell. A urea–oxygen cell which gives about 3 mA cm^{-2} at 0.4 V. has been constructed. Such a power output is unlikely ever to be commercially significant.

One interesting possible application of a biological fuel cell is to provide a miniature pacemaker unit for controlling a patient's heart which has been incapacitated in some way. In theory, such a cell could work from carbohydrates and oxygen found naturally in body fluids thus needing no external supplies, and not—as in present day pacemaker units—require any replacement of electrode or electrolyte materials. Some quite satisfactory experiments have been carried out with *in vitro* arrangements.

Inorganic redox fuel cells

The systems considered in the previous section are really examples of organic or bio-organic redox systems; clearly the more conventional inorganic oxidation reduction systems could also be used. In fact the essence of this type of cell is that the electrode reactions (both 'fuel' and 'oxidant') are chosen for their electrochemical characteristics (that is, high potential and low overpotential) rather than their easy or cheap production. This really means that any redox fuel cell must have either a built in or a closely associated system for regenerating both constituents. Such a system is likely to use oxygen (or air) to regenerate the oxidant, and either hydrogen or carbon (or some reducing agent derived from carbon) to regenerate the 'fuel'.

Materials are generally chosen so that the combination of their standard redox (or electrode) potentials is about the same as the hydrogen–oxygen system (1.2 V). Some suitable systems can be seen from table 3.1 and these include TiO^{2+}/Ti^{3+}, Cu^{2+}/Cu, Hg_2^{2+}/Hg, Sn^{4+}/Sn^{2+} and SO_4^{2-}/SO_2 for the fuel side and Br_2/Br^-, Cl_2/Cl^- and NO_3^-/NO for the oxidant side.

Some examples of electrode systems that have been used are as follows.

(1) The TiO^{2+}/Ti^{3+} fuel system

operation: $Ti^{3+} + H_2O \rightarrow TiO^{2+} + 2H^+ + e^-$

regeneration: $2TiO^{2+} + 2H^+ + H_2 \rightarrow 2Ti^{3+} + 2H_2O$

(2) The Sn^{4+}/Sn^{2+} fuel system

operation: $Sn^{2+} \rightarrow Sn^{4+} + 2e^-$

regeneration: $2Sn^{4+} + 2H_2O + C \rightarrow 2Sn^{2+} + 4H^+ + CO_2$

(3) The Br_2/Br^- oxidant system

operation: $Br_2 + 2e^- \rightarrow 2Br^-$

regeneration: $4Br^- + 4H^+ + O_2 \rightarrow 2Br_2 + 2H_2O$

(4) The HNO_3/NO oxidant system

operation: $4H^+ + NO_3^- + 3e^- \rightarrow 2H_2O + NO$

regeneration: $2NO + O_2 \rightarrow 2NO_2$ followed by

$3NO_2 + H_2O \rightarrow 2HNO_3 + NO$

(regeneration in this case is not complete).

Although in principle this type of cell should be quite satisfactory, in general experimental cells have not worked very well, largely because the separating membrane necessary cannot sustain a very high current density.

Regenerative cells

One of the characteristics of the redox type of cell just considered is the regeneration of materials used at both electrodes. Some mention was made of the ideas of mechanical regeneration during the discussion of air depolarised cells on p. 97, but the principle is a general one and worthy of being considered in connection with all types of fuel cells.

The idea is that the reactants (fuel and possibly oxidant) should be regenerated from the products and recycled, usually by a process taking place outside the cell itself. Some discussion of electrical regeneration, in particular, is more appropriate in chapter 10 where applications of fuel cells are considered, since the 'storage' of electrical energy can be effected by this means. There are, however, several other types of regeneration that will be mentioned here.

Thermal regeneration can be used when the product dissociates on heating to give the reactants again. The two reactants should be readily separable and the process should not require heating above about 1000 °C. Such a reaction will of course be subject to the Carnot cycle efficiency maximum (see chapter 2). The obvious reaction

$$2H_2 + O_2 \rightarrow 2H_2O$$

is not very suitable but a cell has been constructed to use the reaction

$$2Li + H_2 \rightarrow 2LiH$$

The cell operated at 600 °C and had lithium and hydrogen electrodes in a fused lithium–chloride lithium–fluoride electrolyte. The decomposition is carried out at 900 °C.

Other types of regeneration that have been suggested or tried involve conversion of products to reactants by photochemical or radiochemical processes.

Alkali metal–halogen and related cells

Table 3.1 indicates that the highest emf to be obtained theoretically from any cell will probably be from the combination of an alkali metal or alkaline earth metal electrode and a halogen electrode, since these are the simplest systems with the largest negative and positive standard electrode potentials respectively. Some of the problems associated with the possible use of alkali metal electrodes have been discussed previously where a sodium amalgam system was found to be feasible. Halogen electrodes are nearly as difficult to set up, since these materials are also very reactive. For reasons of cost and ease of handling, as well as reactivity, it is probable that only chlorine would prove suitable, unless a regenerative or secondary system could be set up. The difficulties of cell construction with such violently electropositive and electronegative materials present in close proximity are likely to be very great and this is probably the reason why lithium–chlorine cells with, say, a molten lithium–chloride electrolyte have not been developed very extensively, although such a project has attracted considerable interest.

A related cell that has been investigated very thoroughly and which is now in the process of commercial testing is the sodium–sulphur secondary device, which makes use of the reaction

$$2Na(\ell) + S(\ell) \rightarrow Na_2S(\ell)$$

Operational temperatures are quite high, to maintain the constituents in the liquid state, and the usual problems of construction and gasket design have arisen. The system is completely self contained and used strictly as a storage battery (having excellent capacity: mass and capacity: volume ratios) and thus is only rather distantly related to true fuel cells.

CHAPTER 10

FUEL CELL OPERATION

Consideration of the various kinds of fuel cells suggests that it is unlikely that a working cell could be developed that would give a potential difference greater than about 1 V. Most, if not all, practical applications require potential differences considerably higher than this. Thus it is necessary to consider the methods of grouping cells together to form *batteries.* In this chapter these considerations will be outlined in a general way and the problems highlighted. Obviously, many of the difficulties will be economic ones, stemming from the cost of providing the most scientifically suitable arrangements, and so a general discussion of fuel cell economics will be found in chapter 12.

To some extent we have already reviewed problems arising from the operation and the construction of the various types of single cell and, obviously, very similar points will arise when cells are grouped together. We may conveniently list the sources of difficulty here: the electrical linkages between electrodes; the circulation of electrolyte and its regeneration if necessary; the supply of fuel and oxidant to the appropriate electrodes; the removal of waste heat or the provision of heating; and the materials most suitable for construction of the complete system. These are the significant points and we need say nothing further about choice of electrolyte or electrode material since these are considerations of cell design rather than of battery design and as such have been extensively dealt with already.

We must be aware that there may be conflicting requirements, particularly since in almost all applications the power: mass and power: volume ratios should be kept as high as possible. Thus the desire for a compact arrangement may militate against free access of fuel to the electrode or easy circulation of electrolyte. Another possibility is that the lightest materials of construction may be either expensive or mechanically unsound, allowing leakages of current or material.

Supply of fuel and oxidants to the battery

First let us consider the case when the fuel is supplied in the gaseous state. Clearly, similar considerations will apply to the oxidant which is almost invariably air or oxygen. The gas can be either passed successively through the electrode of each cell in turn, which may be called by analogy with electrical systems 'series' operation, or it can be fed simultaneously to the electrodes of all cells, which would, by the same analogy, be termed 'parallel' operation. In some circumstances it may be possible to use a system which is a combination of these two, for example, supplying gas in parallel to blocks of cells within which series distribution is used.

The parallel arrangement is simple to construct and results in little pressure drop, but it may be difficult to ensure uniform distribution to all cells. If air is used as an oxidant, then parallel distribution is practically essential since a high rate of gas flow is necessary; similar considerations would apply to the method of supplying any fuel gas containing appreciable quantities of inert material. Series distribution is often simpler to arrange but inevitably means that each cell receives a slightly lower pressure of gas than the one before it. This means that there must be a limit to the number of cells that can be served in series; a limit is also set by the possible need to have a certain gas flow rate for satisfactory operation. However the passing of gas from one electrode to the next is likely to result in very efficient use of fuel or oxidant.

Liquid fuels can be used of course; at high temperatures of operation they will almost certainly vaporise or 'crack' before coming into the cells and therefore may be treated as gaseous fuels. At low temperatures they are usually dissolved in the electrolyte outside the cell, possibly by some sort of injection process, and the electrolyte is then circulated to the cells in parallel.

Removal of products

Products which need to be removed include water, carbon dioxide and nitrogen (from air or such fuels as hydrazine). Heat may also need to be 'removed'. The use of high temperatures reduces the problem here since excess heat is rarely encountered and any water produced is in the vapour phase and therefore more easily removed. At low temperatures this is, of course, not the case and the water produced dilutes the electrolyte which eventually becomes too weak to support continued efficient operation, although as mentioned in chapter 5 this has been found to be rather less severe an effect than might have been thought. The electrolyte can be restored to a higher concentration by evaporation outside the battery and recirculation. Unfortunately, this may use rather a large proportion of the useful energy of the cell, unless the waste heat can be employed.

Another method of removing water is to have a high rate of gas flow to the electrodes so that the water is carried off as vapour. If an alkaline electrolyte is used in the cells then the water is produced at the fuel electrode, whereas with an acid electrolyte it is produced at the oxidant electrode. Thus the choice of electrolyte determines which of the two gas flows can be used for water removal. If air is used as an oxidant then it is a comparatively straightforward matter to vent the water carrying waste gas straight to the atmosphere. Otherwise it may be necessary to recycle the feed gas over a device to condense out the water, for example, if pure hydrogen is the fuel, since any simple venting procedure would be too wasteful.

Most of the problems about carbon dioxide production and removal have been discussed in earlier chapters, and it is only necessary to summarise the position: there are no difficulties with acid electrolytes since the gas just bubbles off, but aqueous alkaline electrolytes represent an intractable problem. Despite trials of ion exchange systems, electrodialysis and precipitation as calcium carbonate, no completely acceptable answer has been found.

There are several well-established possibilities for removal of waste heat, that is heat produced as a side effect of the electrochemical conversion reaction, equal to $T\Delta S$ in thermodynamic terms for a theoretical cell working isothermally at open circuit potential (emf). These are: external cooling of the battery itself; circulation of a coolant through a system of pipes or interconnecting spaces and galleries within the battery; circulation of the electrode gases through an external cooler; and circulation of the electrolyte through an external cooler. The last two in this list can be combined with devices to remove waste water, as already explained, which will itself, of course, remove heat (as the latent heat of condensation of water vapour). External cooling of the battery itself is usually only feasible for medium or high temperature systems, and is in fact probably not needed for high temperature battery operation where heat may actually have to be supplied to the system to maintain the working temperature.

An interesting proposal has been made to use the waste heat to reform hydrocarbon fuel endothermically before its electrochemical reaction. Such an idea would require very careful balancing of the working cell potential difference and the fuel gas flow.

Electrical arrangements

Apart from the choice of series or parallel electrical connections the problem is essentially one of minimising the total internal resistance and reducing the possibility of electrical short circuits. A contribution to internal resistance is made by the electrolyte and this has been discussed in terms of ohmic polarisation in chapter 4 and elsewhere. There is also a contribution from the electrodes which are not always very highly conducting, and their efficiency will depend on size and probably also on shape. There are two arrangements of neighbouring cells possible, one a *bipolar* system where fuel and oxidant electrodes alternate and are separated by membranes impermeable to gases, and the other a *homopolar* scheme where fuel or oxidant is supplied to two adjacent electrodes at once. Both are shown diagrammatically in figure 10.1 from which it can be clearly seen that the electrical connections for the bipolar arrangement may be by far the simpler of the two types while the supply of reactants will be more complicated.

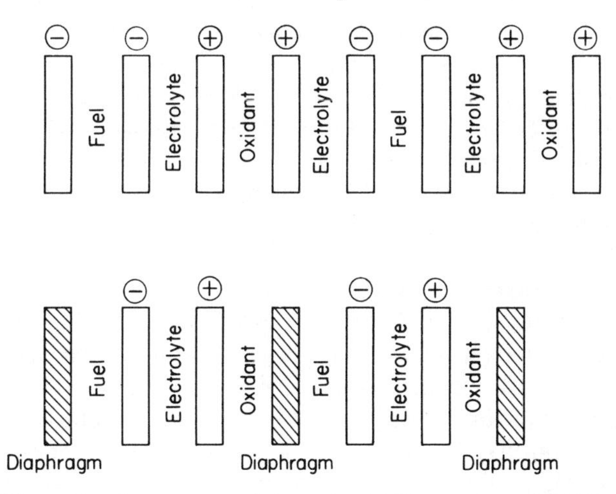

Figure 10.1 Arrangement of electrodes. (a) Homopolar and (b) Bipolar

It has already been seen that electrolyte circulation is frequently employed in fuel cell batteries for water and heat removal and this may lead to an electrical complication; so-called parasitic or shunt currents can occur between one cell and another and reduce the efficiency of the battery, produce additional heat, and cause gassing by electrolysis. Such currents are reduced by provision of long and narrow connecting ports between cells; unfortunately, these characteristics are exactly the opposite of those most desirable for effective electrolyte circulation.

If electrolyte is not circulated, as in some of the ion exchange and capillary cells described in chapter 5, then there is of course no power loss of this sort.

Materials for battery construction

The choice of materials for containing the electrode assembly and electrolyte depends primarily on temperature of operation and, to a lesser extent, on the nature of the electrolyte. At low temperatures—that is, below about 80 °C—various types of polymeric ('plastic') material (for example, fluorinated polymers and polymethacrylates) are entirely suitable. Above this sort of temperature metals must be employed at least for part of the structure, and the range available becomes increasingly restricted as temperatures of operation get higher because of the considerations of melting point and risk of corrosion by the electrolyte. The selection of particular metals depends on their resistance to acid or alkaline electrolytes; for example, some kinds of stainless steels can be used for either variety of electrolyte, but nickel and silver are not stable in acidic solutions. In any case it seems improbable that such expensive metals would be used.

Whatever materials are used, there are bound to be further problems in finding the right substances for manufacture of insulating gaskets and other joining devices. These difficulties increase substantially for batteries operating at high pressures and high temperatures (when, in any case, leaks are more likely) and may be the most significant factor in construction of such batteries. Careful attention to design and selection of materials for the various kinds of peripheral equipment, such as pumps, condensers and compressors, is also important.

Production and purification of fuel

Although cells using several different kinds of fuel have been described in the preceding chapters, it is clear that the simplest of all, hydrogen, is likely to provide the most efficient and successful operation. Unfortunately, hydrogen is a gas of low density which is not easily liquified, and storage and transportation may be inconvenient requiring power for compression or liquefaction, and bulky and heavy containers. However, as was mentioned in chapter 1, it is often possible to produce hydrogen *in situ* from other fuels more readily handled and, as we have seen in chapter 8, such a cracking or reforming process seems to be the first stage in the electrochemical oxidation of hydrocarbons at high temperatures.

Hydrogen has been made from methanol

$$CH_3OH + H_2O \rightarrow 3H_2 + CO_2 \qquad \text{(reforming)}$$

or

$$CH_3OH \rightarrow CO + 2H_2 \qquad \text{(cracking)}$$

from ammonia by cracking

$$2NH_3 \rightarrow N_2 + 3H_2$$

from hydrocarbons by steam reforming as described in chapter 1, from the reaction of iron with steam

$$3Fe + 4H_2O \rightarrow Fe_3O_4 + 4H_2$$

from the water gas reaction

$$C + H_2O \rightarrow CO + H_2$$
$$CO + H_2O \rightarrow CO_2 + H_2$$

as a by-product from the catalytic reforming of hydrocarbons, and by electrolysis of water. The last process is really only useful when the combination of electrolyser and fuel cell forms a device for storage of electrical power. Hydrogen has also been produced by hydrolysis of sodium borohydride for a special application

$$NaBH_4 + 2H_2O \rightarrow NaBO_2 + 4H_2$$

since the mass of hydrogen produced for a given mass of the borohydride is very high. It may be that some other metal hydrides could be similarly used either as primary sources or as storage devices.

The impurities commonly found in hydrogen which may be troublesome include nitrogen (from ammonia cracking, for example), carbon monoxide (which poisons some catalysts), carbon dioxide (which reacts with alkaline electrolytes) and sulphur compounds (which poison nearly all catalysts). All impurities can be removed by diffusion through palladium or palladium–silver, though these membranes are expensive and cannot be used for gases containing certain halogen and sulphur compounds.

Carbon monoxide if present is usually first oxidised in carbon dioxide by the process often known as the water gas shift reaction

$$CO + H_2O \rightarrow CO_2 + H_2$$

over a suitable catalyst. Carbon dioxide is removed by 'scrubbing' with alkaline solutions such as ethanolamine or aqueous potassium hydroxide, or even water. Some sulphur compounds can be removed by this means also, or can be converted to hydrogen sulphide or sulphur dioxide first.

CHAPTER 11

APPLICATIONS OF FUEL CELL SYSTEMS

In our survey of the possible applications of fuel cells to the provision of power for various operations, one way of conveniently dividing up the field would be to consider the different categories of power output requirements. For example, high power for industrial applications, medium power for domestic installations, and low power for certain kinds of vehicle and for use in space. Now, these last two classifications overlap to some extent and consequently it is proposed to deal with types of application—for example, transport—rather than classify installations wholly in terms of power output.

Large scale industrial power generation

In chapters 1 and 2 it was pointed out that, other things being equal, direct electrochemical conversion of energy should be a more efficient process than the conventional means of power production by steam raising and turbo generators, unless the temperature of the steam (or other fluid) used was very high. However, no system of large scale power generation by fuel cells has yet been installed so presumably there must be some other considerations, or the theoretical efficiencies do not each bear the same relation to the actual efficiencies. It is true of course that modern power station technology has reached an advanced stage of development, and whereas at the turn of the century a coal fired steam engine only operated at about 10 per cent efficiency, present day equipment approaches 40 per cent and gas or diesel powered installations may be even better. Moreover other methods of power generation are now worthy of serious consideration; apart from using nuclear reactions to raise steam for the conventional alternators, there is the possibility of developing a magnetohydrodynamic method of energy conversion. This technique uses a kind of heat engine with 'super-hot' flame gases as the working fluid, which is made electrically conducting by ionisation and then passed at high speed through a magnetic field, thus producing an electric current between electrodes suitably placed nearby. The highest temperature reached by the working fluid would be about 2500 °C thus probably increasing the Carnot cycle efficiency. Because of the way in which the energy conversion takes place it is possible to minimise the contact of the hot fluid with the containing vessel and hence reduce the need for costly heat resistant materials of construction. The most favourable competitor is likely to be the gas turbine generator, working at about 850 °C. This is the highest temperature at which turbine blades will have a useful working life. For any of these devices a conversion efficiency of 40–45 per cent is probably the highest obtainable.

A fuel cell installation to give high power, either as a relief power source (for peak hour operation) or for remote areas, must clearly have an efficiency higher than this figure, since it is unlikely to have a very low capital cost (see chapter 12) and thus would not be justified on economic grounds. It is, of course, fair to say that for the continuous operation envisaged for some of these installations, the fuel costs may anyway be more significant than the capital costs.

There is some distinction in requirements between the continuously operating generator and the emergency or peak relief arrangement although both must have the ability to supply short duration peak power demands. The emergency system must have instantaneous starting characteristics and immediate response to variation of load, whereas the most important factor for the other variety is the efficiency of energy conversion.

The requirements seem to be that the oxidant used must be air, since the expense of oxygen production is too great; that the fuel used must be a mixture of hydrogen and carbon monoxide, since reforming of hydrocarbons (for example, from natural gas) to produce this mixture is probably economic, while making pure hydrogen is not; that the power density should be high for cheap long lasting electrodes and other equipment; and that high pressure operation should not be required, since construction and operation costs for the large scale system would be too great. In order to maintain a high working efficiency it is also considered desirable to utilise the waste heat produced by the cell (if the electrochemical conversion is 50 per cent, say, then the remaining 50 per cent of chemical energy supplied becomes waste heat) and this can best be done by using it to raise steam which could then drive a conventional generator. Such an addition would cause an increase in the capital cost per unit of power, of course, but may otherwise be advantageous. Moreover, this heat–electrical energy conversion is inefficient unless carried out at about 600 °C which implies fuel cell operation at that temperature or above.

A review of all these requirements suggests that the molten electrolyte cell operating at a high temperature with nickel or nickel oxide electrodes described in chapter 8, will probably be the most suitable, although the low temperature system might offer advantages for emergency power generation where overall efficiency is less important but immediate starting is essential.

The individual cells would be as large as possible and formed into batteries with the bipolar electrode arrangement described in chapter 10. This minimises the cost of gaskets, supporting framework and electrical connections. The complete structure would either be made of heat resistant steel with ceramic gaskets or entirely of ceramic materials. One suggested arrangement is to have hundred-cell batteries, each cell having electrodes of area about 300 cm^2. These batteries would produce about 125 kW at 62.5 V and could, if desired, be grouped in larger sets. Waste heat is removed largely by circulation of the oxidant, air. This heat is used to preheat both fuel and air and also to raise steam in the way described earlier. Some of the preheating can be done by recirculating the fuel gases which process is, in any case, desirable to minimise carbon deposition around and in the electrodes and to introduce carbon dioxide into the oxidant stream (for reasons discussed in chapter 8).

The economics of such an installation are described in the next chapter, but it is important to realise that additional capital and running costs are likely to be incurred by the provision of an inverter—a device to convert a direct current supply into alternating current. It is almost essential to provide ac for large scale power supplies, largely because of the ease with which it can be transformed to lower and higher potentials for transmission or for different kinds of electrical application.

Large scale energy storage

It is well known that conventional methods of power generation are most efficient when providing a constant power supply, irrespective of time of day or of demand. Demand varies very considerably both within one day and also from one day to another, as well as being rather lower in the summer months (this may not be true in all countries since, for example, peak power in New York City is required during the high summer when air conditioning plants are running most of the day and night). Thus storage of electrical energy to even out demand is desirable; the only practicable method of storage is to use some reversible conversion into another form of energy. An example of a method that has been used is the pumped storage scheme, where electrical energy produced during a period of low demand is used to pump water into a high level reservoir; the water can then run down through turbines to generate electrical power again when demand is high. Electrochemical energy conversion provides another possible method; an example was quoted in chapter 7 where water was electrolysed by power supplied to form hydrogen and oxygen. These gases can then be stored until power is required again, when they are permitted to operate a fuel cell. There are some problems connected with using the same unit both for generation of current and for electrolysis since electrolysis tends to damage the electrodes and there would certainly be expenses connected with control gear, storage of gases and maintenance of electrodes. It would be necessary to rectify incoming ac and use an inverter to deal with the dc produced by the cells, which would probably be of the Bacon high pressure hydrogen–oxygen type.

Power plants for vehicles

In this section we consider the use of fuel cell systems to power road vehicles. Various types of vehicles may be considered, each with different working characteristics. The range will run from heavy lorries and buses for long distance travel, through private cars of various sizes to delivery vehicles either of the milk float type or of the type used as internal works transport (for example, fork lift trucks). Two special varieties are taxis or other small vehicles for transport in cities, and agricultural tractors.

For all vehicles two characteristics seem essential: a quick start and a short time of response to variation of load. Overload capacity is also important. Critical factors are undoubtedly the power: mass and power: volume ratios of the power plant, its peripheral equipment and fuel containers. Fuel cells offer very considerable advantages in this particular field since they generally operate with no emission of toxic or offensive products and with a minimum of noise and vibration. Moreover, dc electric motors are usually thought to be more suitable for road traction (or for rail traction, for that matter—see p. 114) than ac motors, and therefore the fuel cell power plant is very appropriate, although it is true that recent developments have tended to make the ac motor more comparable in suitability.

The requirement of immediate starting (which the internal combustion engine has) seems to rule out medium and high temperature fuel batteries unless starting is effected from some other system (for example, ordinary storage batteries), changeover being carried out when fuel cell working temperature is reached. Such a scheme does not sound very practical and consequently most attention has been given to the low temperature systems. For example, cells using aqueous phosphoric acid at moderately low temperatures with hydrocarbons as fuel have been studied, although the type using methanol dissolved in the electrolyte may also be worthy of attention. A disadvantage of all these systems is their reliance on platinum and other noble metals, all of which are expensive.

Several investigations have shown that for almost all sizes of private car used in several different ways the fuel cell shows a considerable decrease of fuel cost per mile. This decrease is least for the largest vehicles. Maintenance costs are also likely to be lower than for the petrol driven car used today but, unfortunately, capital costs of all fuel cell systems are much greater than those of the internal combustion engine, as explained in the next chapter. However, with the price of oil fuels having risen by 100 per cent in the past year and the prospect of further increases in the future, any more efficient use of these fuels (as in the operation of fuel cells) could become more attractive.

For many years small delivery trucks have been operated by storage batteries; these are vehicles having intensive operation over relatively short periods of the day but rarely travelling long distances. Understandably replacement of the storage batteries by fuel batteries has been tried; performance is very satisfactory but again capital cost is likely to be the disadvantage. Further consideration is given to these types of light vehicle, particularly those in industrial use, in the next chapter.

Recent research on electric cars seems to have been concentrated on a different kind of cell, the alkali metal–halogen storage battery mentioned briefly in chapter 9.

Fuel cell powered railway locomotives

Battery powered locomotives and multiple unit railcars have been extensively tried out over a period of many years, and have operated with success in countries where the cost of diesel fuel is high. Again, it seems likely that the fuel battery system could replace the storage batteries satisfactorily and this is particularly true for rail traction since, although load variations are large, maximum power is only required for relatively short periods of time, and this characteristic is very favourable to the operation of fuel cells. Weight and volume limitations (which tend to be rather restrictive on railway vehicles) may prove important in the use of fuel cells, although, apart from that, both low temperature methanol cells and Bacon type cells (possibly with built in reformers) seem possibilities. It appears likely that the lighter, shunting type of locomotive may prove a better application than the larger and heavier main line version.

Submarines

At the present time submarines are either powered by a combination of diesel electric generation and large storage batteries or by nuclear power, presumably via thermal and electrical energy conversion. A low temperature hydrogen–oxygen fuel battery system would be a good replacement for the first of these varieties and calculations have shown that such submarines would have an operating time up to thirty times greater than the conventional battery powered vessels, largely because of better space utilisation. There would be no problems of noise, or of pollution, and several sizes of installation can be envisaged. For such essentially military applications the ordinary economic considerations are clearly irrelevant. A military application somewhat comparable with the fuel cell powered submarine is illustrated in figure 11.1, which shows a naval craft propelled by a fuel battery.

Domestic power

A very high proportion of the cost of supplying electrical power by conventional means to all consumers is the cost of transmission (as high as 60 per cent of power supply costs). Moreover, this proportion may well increase as countries become more industrialised and the demand for power becomes greater and the siting of power stations becomes more difficult; the capital costs of the very large stations now favoured (2–4 GW, for example) are enormous. It has been suggested that the complex national grid system used

Figure 11.1 Naval craft. This boat was designed by the Admiralty Materials Laboratory; it is powered by a 700 W fuel battery capable of delivering 30 A at 24 V. Gas supplies are provided for in small cylinders located in the stern of the boat. Refuelling can be accomplished in a few minutes. An important advantage from the military standpoint is the silent operation of the unit

so far in Britain might be supplemented by small local power plants of the fuel battery type for particular applications, and obviously certain low power electrical installations in remote places might well be supplied by fuel cells. The kind of applications that might be best served in this way include marine beacons, railway signals, remote telephone exchanges, isolated lighting, radio transmission and even power in the home. There are military versions of some of these applications, too, and here additional advantages can be mentioned: fuel cells are quiet in operation and difficult to detect, particularly if they work at low temperatures; they are reliable and have a better capacity than conventional storage batteries; and do not need a bulky and noisy generator set for charging. Two hydrogen fuel batteries of types suitable for low-power application are illustrated in figures 11.2 and 11.3.

As long ago as the eighteen nineties it was suggested that houses might have individual fuel cell plants for providing the low power necessary for domestic electrical apparatus, and that liquid or gaseous fuel would be supplied to domestic consumers rather than electricity. From the earlier comments we can see the superficial attraction of such an idea. The fuel supplied could be natural gas (although this may require reforming) or hydrogen. Hydrogen could be delivered in cylinders but it may be possible in the future to devise a distribution network for this gas too (see chapter 13).

Figure 11.2 Prototype of compact stack-built hydrogen–oxygen fuel battery

Figure 11.3 Hydrogen–air $1\frac{1}{2}$ kW fuel cell. Stack-built electrodes, intended for the Ministry of Defence

The most convenient type of cell to be used in this kind of application would appear to be the low temperature hydrogen–oxygen cell, since it is likely to require least attention and maintenance and can supply power very quickly on demand. However, such an installation has never been made; the capital cost would undoubtedly be beyond the reach of most householders and, at present, the price of hydrogen for this sort of application is not economic compared with a mains electricity supply. Details of economic calculations will be found in the next chapter.

An unusual kind of total power system was that constructed in about 1969 for supplying an 'underwater house' intended as a base for exploration of the sea bed. This was a 50 W hydrogen–oxygen unit and was designed to provide power at 12 V for 1 week, chiefly to run fluorescent lights and an air purification system. The low potential difference was chosen deliberately to minimise any dangers from electrical leakage through the sea water. The tests were successful but the unit appears not to have been obviously superior to an ordinary storage battery arrangement.

Two small-scale applications are illustrated in figures 11.4 and 11.5; the first is a navigation beacon for shipping, the second a runway light for aircraft. Both can be operated essentially without attention over long periods.

Fuel cells in space

So far in this chapter all the applications described have been purely speculative; none of them has actually been set up and operated under any but experimental conditions. The one environment where fuel cells have been

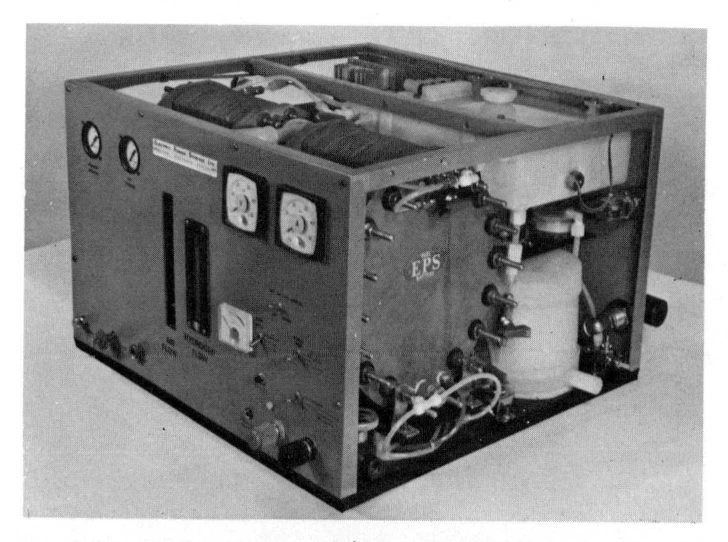

Figure 11.4 12 V, 5 A fuel cell intended for Trinity House for use on beacons. It could run for up to 12 months unattended

Figure 11.5 300 W, 12 V hydrogen–oxygen fuel battery which could be buried in airfield runways to power landing lights

installed and operated under conditions where success was essential has been outer space; space exploration has involved the sending up into the regions outside the earth's atmosphere of various devices to collect scientific information and transmit it back to earth. Electrical power has been required for operation of data collection and transmission apparatus and sometimes for other purposes, particularly in those space devices which have carried men.

Fuel cells were first developed for use in space vessels because although the low power devices mentioned in the previous paragraph could be supplied by some form of primary electrochemical battery, most of the batteries generally used had comparatively short lifetimes (up to about 30 hours for a cell having a power: mass ratio of about 550 W kg^{-1}) whereas the lifetime required was likely to be much longer, perhaps 200–300 hours for the same power: mass ratio. So, particularly where the power required was greater than about 200 W, the fuel cell was highly suitable, especially since the normal economic pressures of commercial development were of little account here.

The only cells used up to the present time, so far as is known, have been of the hydrogen–oxygen type operated on supplies of these liquified gases which were carried for use in rocket operation. Consideration has been given to the use of regeneration or secondary cells taking their primary power from the sun; these would of course solve the problem of how to carry enough fuel for the long voyages into space which will presumably become more likely in the future. Another way that has been suggested for overcoming this problem is to construct a biochemical fuel cell of the type

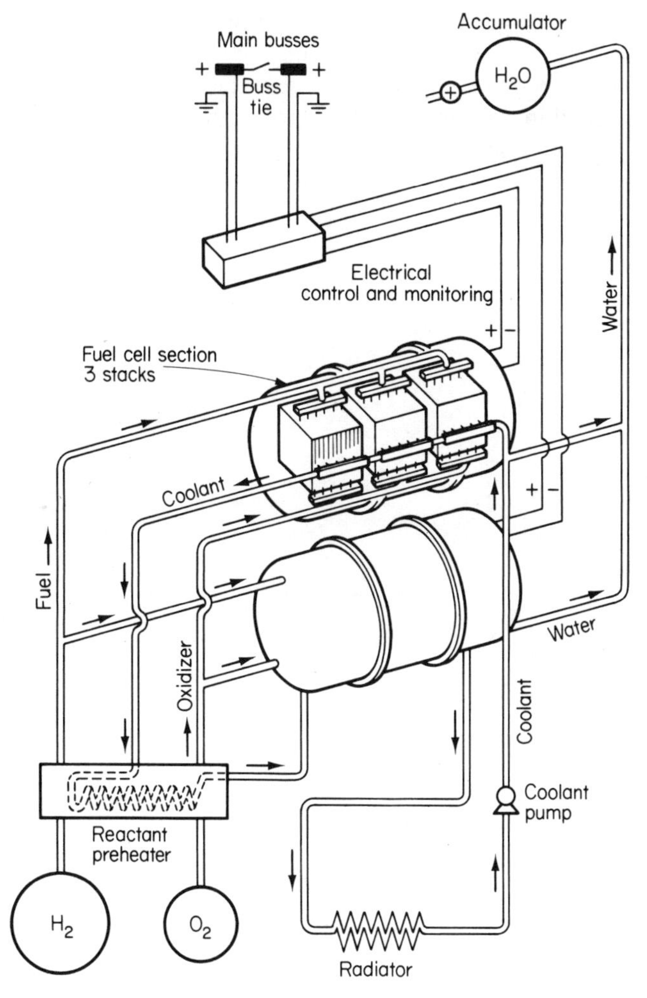

Figure 11.6 A schematic diagram of the Gemini fuel cell

described in chapter 9, where biological waste (for example, astronauts' urine) is the source of the chemical energy. Such a cell has, of course, the additional advantage that human waste products can in principle be converted into potable water at the same time. However, as was explained in chapter 9, apart from the technological problems in setting up such cells the power obtained from them is unsatisfactorily low.

Space vessels are likely to spend much of their voyages in regions of zero gravity, and therefore cells without free flowing electrolyte are likely to be necessary. The membrane and capillary varieties of cell were discussed in chapter 5 and the solid electrolyte type in chapter 8. A membrane cell was used in the Gemini spacecraft and a schematic diagram of the installation is shown in figure 11.6. For the Apollo spacecraft a modification of the Bacon cell described in chapter 7 was used; the concentration of potassium hydroxide solution was increased in order to allow operation at lower pressures and temperatures and also to reduce the mobility of the electrolyte in weightless conditions.

CHAPTER 12

FUEL CELL ECONOMICS

If we are to understand why fuel cells have been used only very little so far and to predict what will be their most favourable application in the future, then we must know something about the economics of their operation and compare this with the economics of other forms of energy conversion. In this chapter we shall try to evaluate the economic factors of fuel cell operation and make some attempt at comparison with its competitors, although this will not be very satisfactory since any costs chosen at the time of writing may be quite wrong by the time this book is read. In fact, of course, the whole field of economics is subject to this uncertainty and this must be borne in mind throughout the chapter.

After an initial consideration of the general economics of fuel cell operation, it will be most convenient to divide our examination into three parts concerned with low, medium, and high power batteries, and these will correspond more or less with the different kinds of application described in the last chapter.

The basic division of economic factors

There are three elements concerned in the economic analysis of fuel cell operation.

(1) Fixed capital expenses.
(2) Fixed working expenses.
(3) Proportional working expenses.

They may be explained further as follows.

The *fixed capital expenses* include:

(1) The interest on the capital expended on the equipment and building.
(2) The amortisation (or writing off) of this equipment.
(3) The amortisation of any setting up expenses, and possibly of design and research costs attributed to the project.
(4) Insurance.
(5) Rates and taxes on equipment, buildings and site.
(6) Some allowance for inflation and possible devaluation.

All of these charges are constant and do not normally vary from year to to year. Moreover they are not obviously related to the power output of the installation.

The *fixed working expenses* include:

(1) Wages and salaries of maintenance and operating staff.

(2) Supplies for maintenance (not for operation).

(3) Fuel for maintenance (not for operation).

These costs cover the maintenance of the power plant, whether it is used or not.

The *proportional working expenses* are essentially only the cost of fuel and oxidant, if used, although they may also include any additional costs required to produce maximum power.

The cost of electrochemical energy conversion in a fuel cell

For a fuel cell working at maximum power, P_{max}, the values of the three elements described in the previous section can be fairly readily obtained. The fixed capital expenses element, R, is given by

$$R = \frac{C_{max}(x + 1/L)}{W}$$

where C_{max} is the investment cost per unit power of P_{max}, L is the lifetime of the installation, x is the rate of interest payable on the invested capital, and W is the fraction of the year when the cell is in operation.

The fixed working expenses element, F, is given by the cost of operation of the whole installation in unit time divided by the maximum power produced, P_{max}.

The proportional working expenses element, K, is given by a simple expression

$$K = \frac{C_{fuel}}{E\eta}$$

where C_{fuel} is the specific cost of fuel used (that is the cost per unit mass of fuel), E is the specific energy produced (the energy produced by the theoretical electrochemical reaction per unit mass of fuel), and η is the working energy efficiency (the ratio of work done to energy put in, as defined in chapter 2) at maximum power.

Some simple conclusions are possible: as the lifetime of the installation increases, so R decreases. R also decreases when the rate of utilisation is high (that is, the working period is a large proportion of the year). K depends inversely on the working efficiency of the plant. The measurement of R, F and K is in cost per unit power per unit time, and the conventional units are therefore pounds (or other appropriate currency) per kW h.

When the installation is not working at maximum power, it may be said to have a *load coefficient*, f, given by

$$f = \frac{P_{actual}}{P_{max}}$$

where P_{actual} is the actual power produced. In these circumstances the new values of R and K are given by the primed quantities

$$R' = R/f$$

$$K' = C_{fuel}/E\eta'$$

where η' is the efficiency of the system for production of power P_{actual}. It seems likely that η' is directly proportional to f, at least when f lies between 0.1 and 0.7.

The element of cost F has two components, the fixed expenses and that part of the maintenance costs which depend on extent of utilisation. However to a first approximation, F will be unaffected by changes in f since the divisor is the power produced, and this of course determines f. Thus the total cost per unit power per unit time is the sum of the elements R',F and K':

$$C = R' + F + K' = F + \frac{C_{max}(x + l/L)}{fW} + \frac{C_{fuel}}{E\eta'}$$

There will be a minimum in C corresponding to a particular value of f, which may not be 1. This value, f_{min}, will be influenced by, among other things, the increase in C_{max} (increases f_{min}), the lengthening of the cell lifetime (increases f_{min} up to a point), the increase in fuel costs (decreases f_{min}), the increase in the utilisation, W(decreases f_{min}), and the increase in the maximum efficiency (increases f_{min}). Further amendment of C may be necessary if any power is used by peripheral equipment.

Evaluation of the *R* (fixed capital) and *F* (fixed working) elements of costs

Before we move on to consider the different types of application of fuel cell systems, we should perhaps discuss the factors which will determine the values of these three elements of cost defined in the two previous sections. In fact, of course, any estimates of these elements, in particular the R (capital costs) and F (fixed working costs) elements, will be largely speculative since there are little or no data on the commercial operation of fuel cell systems, and factors like utilisation and working life are unlikely to be capable of being accurately forecast. In this section it is assumed that the three cost elements are independent of one another, although this may not be strictly correct.

A distinction may be drawn (as was done in the previous section) between the investment cost per unit of economic power (that is, the power produced at optimum proportional working cost) and the investment cost per unit of maximum power; the relation between these two is likely to depend largely on fuel cost. Overload capacity is important in many applications and therefore must be allowed for.

Low temperature installations, as we have seen, require expensive catalysts as constituents of the electrodes and are likely to run on expensive fuels, possibly produced on site from cheaper materials. However, they do not have structural or corrosion problems; the containing system and associated equipment will not be expensive to construct. It was estimated in 1968 that an installation of this type could be produced at a cost of about £100 kW^{-1} for economic or optimum power production; at that time it was also considered that technological advances gained during a period of mass production of such systems could reduce the cost to about a quarter of this figure, and that the cost for maximum power would be about one half of that. Low temperature batteries of the ion exchange membrane type will be considerably more expensive—up to seven times the cost of systems using aqueous electrolytes—but again this figure is expected to decrease sharply during development in the future. However, this type of installation has a poor overload capacity and the cost per unit of maximum power will always be relatively high. Batteries working with dissolved fuels (for example, methanol or hydrazine) are about 50 per cent more costly to construct and install than those using hydrogen.

High temperature systems have cheap electrodes and use cheap fuel compared with those operating at lower temperatures, but the constructional costs are likely to be expensive and corrosion problems almost certainly increased. Moreover, additional expensive peripheral equipment is likely to be required—start-up heating and pressure control systems, for example. The investment cost is likely to be almost as great as for the ion exchange membrane cells, but since less experience has been gained with this sort of battery system, production is considerably less reliably costed than for low and medium temperature operation.

Medium temperature cells (the Bacon type for example) are comparatively well established and their investment costs were estimated in 1968 to be lowest of all; but because of their more thorough investigation, there is less likely to be significant technological advance in their construction in the future and the cost is not expected to fall so much.

In the long term, then, we might predict the low temperature fuel batteries to have the lowest capital investment costs.

The *F* element (that due to fixed working costs) is generally less significant than the other two; maintenance of fuel cell systems is not usually very arduous or expensive and consequently the costs of supplies, wages and salaries associated with this activity do not reach a very high level compared with other contributions.

Evaluation of the *K* element of cost (fuel expenses)

Costs other than those of fuel contained in this element are probably only a small part of it, and moreover can probably be regarded as directly proportional to the fuel cost. Thus we need only consider the cost of fuels here.

Since almost all fuels used in fuel cell installations will be obtained from natural gas or oil, prediction of their cost in the future is subject to more than the usual uncertainty. However, some current prices or estimates are

Table 12.1 Some fuel prices

	Price/p MJ^{-1}
Hydrogen (in cylinders)	0.028
(from plant)	0.087
Hydrazine	4.0
Natural gas (methane)	0.012
Reformer gas	0.04
Ammonia	0.016
Methanol (present process)	0.17
(from oil)	0.071

These figures are approximate and of course may vary widely according to locality and other circumstances. They are intended to show the relative magnitudes.

Prices per energy unit refer to the energy of combustion. Clearly, the price per electrical energy unit would be obtained by multiplying this figure by the energy efficiency of the fuel battery system.

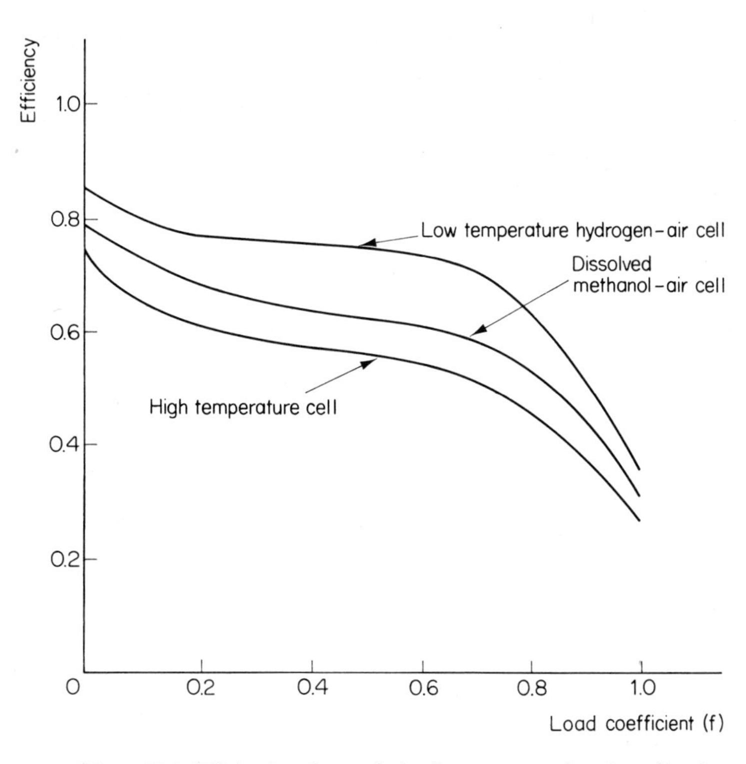

Figure 12.1 Efficiencies of some fuel cell systems as a function of load

contained in table 12.1. In general the cost of any fuel also depends on the quantity required; thus the supply of fuel to a large installation may be relatively cheaper than that to a small one. Some reference is made to this in table 12.1.

Of the fuels commonly suggested for fuel cell operation, hydrogen is the simplest but must be produced either near at hand, or at a remote site from natural gas, oil, or possibly coal. Use of small quantities may be relatively expensive since on site production is likely to be uneconomic, and storage and transportation of such a dangerous material will also be costly not only because of hazards but also because of its low density and the difficulty of liquefaction. Methanol may be more favourable in these circumstances, despite its greater cost for the same power produced. Hydrazine is at present prohibitively expensive but might possibly become cheaper and has other advantages (see chapter 6) which may be significant for particular applications.

The energy efficiency, η, also enters the expression for the K element of cost; this will vary with type of cell, of course, and also with load factor. Some representative efficiencies are shown in figure 12.1.

Low power applications

This category includes all installations of fuel cells yielding less than 10 kW and consequently includes most of the small static applications described in chapter 11. It also includes the power plants for small industrial trucks and similar vehicles, and also the applications of fuel cells for military

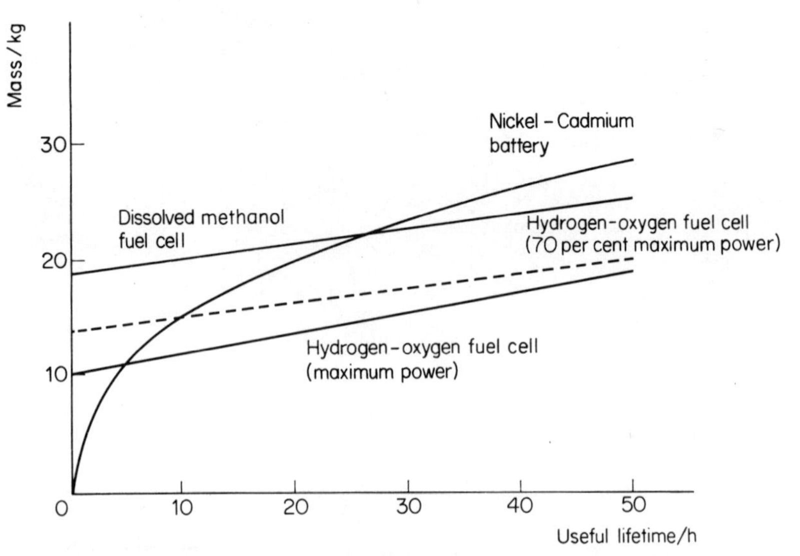

Figure 12.2 Relation between the mass of some batteries and fuel cells, each comprising a 200 kW unit, and the lifetime of their use

purposes and in spacecraft, although these last two will hardly be subject to our economic scrutiny.

For many of these applications the weight of the battery used is significant and its dependence on lifetime is also important. A graph of this dependence for several devices appears in figure 12.2.

A very important potential use of low power fuel cell systems is in the light industrial vehicle, such as the fork lift truck. Such vehicles are not normally required to operate at high speeds and may stand idle for long periods between times of high activity. It has been estimated that a fork lift truck of mass 1 ton (1000 kg) consumes 2 W h on starting, requires 1 W h for movement of 10 m on the level and about 7 W h for lifting 1 ton through 1 m. However overloading of the power supply can often take place. At present this kind of vehicle is frequently powered by storage batteries since charging is readily carried out during the periods of idleness and since the operating costs seem to favour electric operation; in 1967 it was said that the cost of internal combustion engine operation was about 40 p h^{-1} for a 2 ton truck whereas the battery system cost about 23 p h^{-1}. It can reasonably be assumed that 1975 figures would be about double these.

There would be no great difficulty in substituting fuel cells for the conventional storage batteries in such vehicles as fork lift trucks; in fact, it has often been suggested that this application of fuel cell electricity supply will be the first important one other than in spacecraft, and consequently the first truly commercial one.

The power: mass and power: volume ratios for various sources of supply are quite important here, and some representative figures are shown in table 12.2. An exact comparison, however, is not possible since no account is taken of the masses and volumes of fuel containers or other peripheral apparatus (except pumps and circulation equipment), which leads to fuel cells having apparently an obvious superiority over the storage batteries. However, storage batteries would require costly rectifying and charging equipment to be available at a suitable site.

The calculations of cost per unit time referred to earlier have been extended

Table 12.2 Mass: power and volume: power ratios for various energy sources

	Mass: power/kg kW^{-1}	Volume: power/dm^3 kW^{-1}
Internal combustion engine	3	3
lead–acid battery	600	330
nickel–cadmium battery	200	100
low temperature hydrogen fuel cell with liquid electrolyte	15–20	20–30
low temperature dissolved methanol fuel cell	30–45	60–90
low temperature dissolved hydrazine fuel cell	10–15	15–25

to cover various forms of fuel cell powered trucks; the zinc–air and and hydrogen–oxygen systems lead to figures comparable with the internal combustion engine, of about 39 p h^{-1} and 37 p h^{-1} respectively. Dissolved methanol cells provide a figure closer to the lead–acid battery, of about 26 p h^{-1} and the low temperature hydrocarbon cell (not yet properly developed) would have a cost of about 15 p h^{-1}—the lowest of all.

Medium power applications

The range of power covered by this description is between 10 and 150 kW. This range includes small power generators and most forms of road vehicle.

Economic considerations for generators differ according to whether they are required for continuous service or emergency use. Clearly the emergency systems require fast starting and high efficiency but fuel cost is perhaps less important than it is for the units in continuous service. For some emergency uses, such as emergency hospital power, quiet operation and a minimum of pollution are obvious advantages. However their capital cost should preferably be as low as possible and so should any maintenance expenditure, since for much of their life there can effectively be no return on these expenses. In both kinds of generators a high overload capacity is required and this is where the fuel cell system may have a distinct advantage over the most obvious competitor, the diesel engine alternator set, since fuel cells work most efficiently when loading is low compared with their

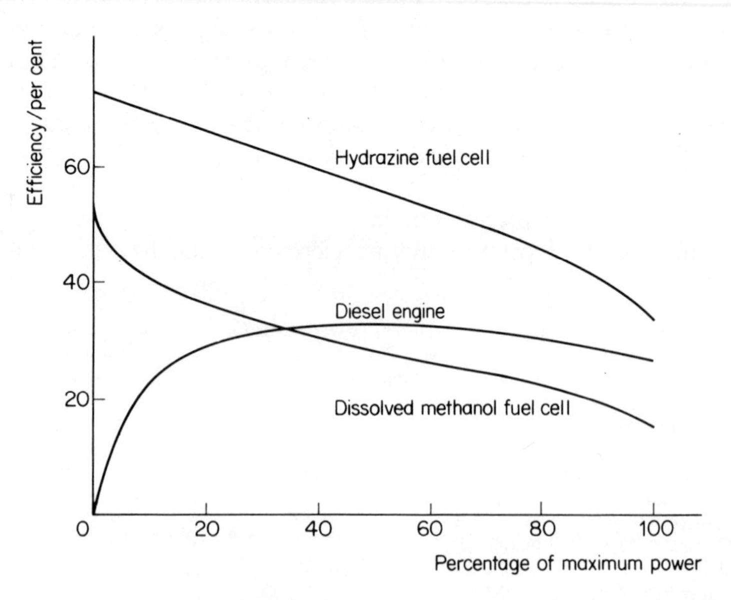

Figure 12.3 Relation between efficiency and percentage of maximum power for the diesel engine and two kinds of fuel cell

maximum power output. This characteristic is illustrated by the graphs for several varieties of power plant shown in figure 12.3.

There are two particular economic effects applicable to these types of generator: the annual operating time is low for standby installations, thus making W small and increasing the R element, and if an inverter is required (to produce ac) then the K element is increased by 1/0.85 at least. This factor assumes an inverter efficiency of 85 per cent. Of course there will be an additional capital cost here as well. Nevertheless, the low temperature fuel cell system does not seem to compare favourably with the conventional arrangement for emergency operation.

Calculations show that continuously operating fuel cell power generators are not as suitable as the conventional systems, since the capital cost is less significant but the fuel cost is likely to be high, although—as we have already said—the fuel cell system is particularly competitive when working at a low load.

Installations where fuel cell energy is to be converted into mechanical work (by means of an electric motor) give the fuel cell systems a slight disadvantage since now the motor has to be included in the efficiency calculations. The graphs for this kind of operation are shown in figure 12.4. Despite this reduction and the unfavourable values of power: mass and power: volume ratios, the fuel cell plus motor arrangement does have one advantage over the internal combustion engine—a higher torque at low speeds—and therefore does not require variable gearing. This means that maximum power requirements may well be lower.

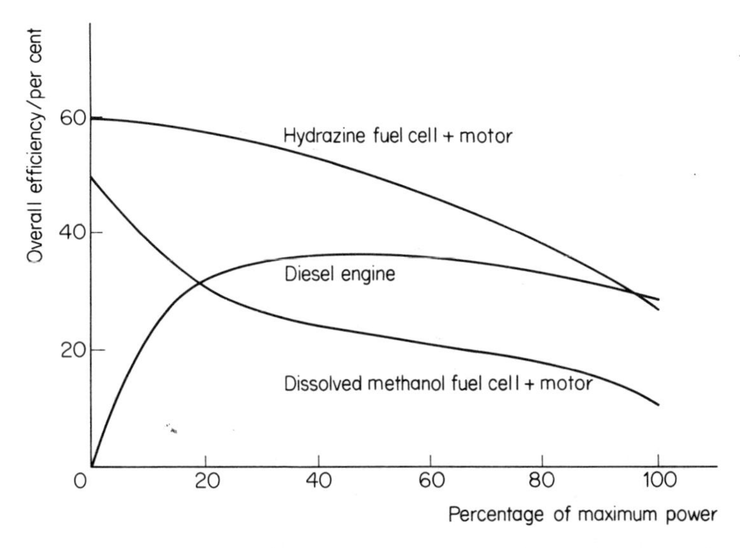

Figure 12.4 Relation between efficiency of conversion into mechanical energy and percentage of maximum power for the diesel engine and two kinds of fuel cell powered electric motors

Table 12.3 R, F and K elements for various types of road vehicle using conventional internal combustion engine

	Taxicab	Private car	Delivery truck	Works vehicle
R	0.219	0.308	0.188	0.312
F	0.308	0.342	0.385	0.385
K	0.512	0.342	0.946	0.915
Total	1.039	0.992	1.519	1.612

The units are pence/kW h
The figures were calculated in 1967 using the US price for petrol with taxes subtracted. 1975 values would be considerably greater but the precise amounts are unclear.

The power plant for road vehicles at present is almost exclusively the internal combustion engine, either in the petrol or the diesel variant. It has many advantages: easy starting, maximum efficiency at maximum power, quick response to load changes and easily handled high energy fuels. The values of the R, F and K elements for these units in some varieties of road vehicle are summarised in table 12.3. Fuel cells for road traction have clearly to compete with these characteristics.

One of the biggest obstacles to the use of fuel cells is their poor power:

Table 12.4 R, F and K elements for various types of road vehicle powered by hydrogen–oxygen and methanol–oxygen fuel cells

Hydrogen–oxygen	Taxicab	Private car	Delivery truck	Works vehicle
R	0.565	0.969	0.396	0.323
F	0.104	0.115	0.127	0.127
K	3.058	3.038	2.746	2.746
Total	3.727	4.122	3.269	3.196
Difference from internal combustion engine (table 12.3)	+ 2.688	+ 3.130	+ 1.750	+ 1.584
Methanol–oxygen				
R	0.677	1.185	0.431	0.323
F	0.104	0.115	0.127	0.127
K	1.058	1.158	1.038	1.058
Total	1.839	2.458	1.596	1.508
Difference from internal combustion engine (table 12.3)	+ 0.800	+ 1.466	+ 0.077	− 0.104

Units are pence/kW h.
Figures were calculated from 1967 estimated prices. 1975 values would clearly be different but the precise variation is uncertain. The comparisons may still be valid.
In the difference figures quoted, positive quantities indicate greater cost for the fuel cell system.

mass ratio compared with conventional systems. It can be calculated that the figure for battery plus motor is about 25–40 kg kW^{-1} for a dissolved fuel–liquid electrolyte type of cell, and although this may be reduced in the future it is still much greater than the internal combustion engine (5–10 kg kW^{-1}).

The results of some calculations are shown in table 12.4, which summarises the values of *R*, *F* and *K* elements for hydrogen and methanol powered low temperature fuel battery systems for various types of road vehicle. As we might guess, the lower the percentage of maximum power used the more favourable is the electric vehicle, and because of their high investment costs the fuel cell powered vehicles are also favoured by high utilisation (as in the taxicab). It will be noted that the methanol cell appears to be a better performer than the hydrogen type for all vehicles considered. Nevertheless, at present there seems no economic reason to replace the well-established petrol and diesel driven road vehicles; only if capital costs or fuel costs decrease more substantially than current predictions suggest will this change become favourable. As we have indicated previously it might be that environmental considerations (lack of noise, no polluting exhaust) in certain circumstances could outweigh the economic ones.

High power applications

Installations requiring a power greater than 150 kW include high power generators, ships and railway locomotives. For all of these applications fuel cost appears to be critical. Table 12.1 gave some costs of fuels likely to be used in fuel cells; it is, of course, subject to the usual uncertainties about prices, particularly their future trends, and it should be pointed out that these costs depend very much on the geographical location where they are required. Waste hydrogen from a reforming plant will clearly be very much cheaper when produced in close proximity to the fuel battery than it would be when made hundreds of miles away.

Large scale power generation appears to be quite favourable because of the superior efficiency of the fuel cell and also because both low and high temperature cells using hydrogen are possible alternatives to the conventional arrangements, provided that only continuously operating power plants are considered. Any intermittent operation would not be well served by the high temperature types of cell. If the same fuel is considered to be used in both conventional and fuel cell installations, then the higher efficiency of the fuel cell mentioned earlier means that for an expensive fuel the fuel cell system has an advantage.

For railway locomotives, calculations suggest that two types of locomotives can be distinguished: the main line engine with lower *R* costs but higher *K* (fuel) expenses than the diesel–electric type usually used, and the shunting engine which shares the same fuel cell characteristics but is more expensive than the main line version when powered by diesel–electric means. These observations support the conclusions of chapter 11.

CHAPTER 13

FUEL CELLS AND THE FUTURE

Despite the numerous possible applications of fuel cell systems described in the last two chapters and despite the impressive advantages and superiority of fuel cells over other kinds of power generation, the only application that has passed beyond the strictly experimental stage is the supply of electrical power in space vehicles. The reasons for this are apparent from the economic considerations outlined in chapter 12 but are reinforced here.

Fuel cells are provided in spacecraft because they are the best devices for the job, quite independently of any analysis of cost. As we have seen, when cost is assessed, fuel cell systems do not yet appear competitive, except perhaps marginally for such applications as emergency power suppliers and light industrial vehicles.

Summary of advantages of fuel cells

From purely theoretical considerations fuel cells ought to provide a more efficient way of utilising the energy of combustion of a fuel than the conventional forms of power generation, except perhaps for the conventional type working at high temperatures. Further, the efficiency of any form of conventional heat engine increases with increasing power output, whereas the opposite is true for a fuel cell, largely since high currents produce high overpotentials. This, of course, as has been pointed out several times in the last chapter, means essentially that overload capacity can be built into the system automatically

The efficiency of a fuel cell appears to be independent of its size and, in general terms, of the number of cells linked to form a battery. However, other considerations may dictate certain minimum desirable configurations. For example, the supply of compressed gases (particularly air) may be uneconomic for a system of less than a certain capacity, and the high temperature type of cell may have too high a rate of heat loss to be self sustaining unless it is more than a certain size.

It has often been claimed that fuel cell power supply is kinder to our environment than the means generally used today, in that the main waste products of fuel cell operation are water (or water vapour) and carbon dioxide, neither of which is generally thought of as a severe pollutant.

However some doubts have been expressed about the effect on the solar energy received by the earth of large quantities of water vapour or carbon dioxide in the upper atmosphere, so this advantage may be illusory. But it must be remembered that ordinary combustion of fuels produces these materials as well as more obviously offensive substances, and moreover does it less efficiently.

Noiseless operation, too, is often quoted as an advantage possessed by fuel cells over other forms of power supply. This is also an environmental consideration and may in certain circumstances override ordinary economic calculations. In general, this has not happened so far because either the economics are so heavily against fuel cells that no user can afford to take account of the environmental advantages, or those applauding the silent and pollution free operation of fuel batteries have been unable to convince the rest of the world that this is worth paying for. The most likely application where these characteristics might be worthwhile is probably the provision of light electric transport in large towns and cities or in industrial complexes, although they might also be significant in the provision of emergency power for hospitals.

For portable operation the mass of the generating equipment is very

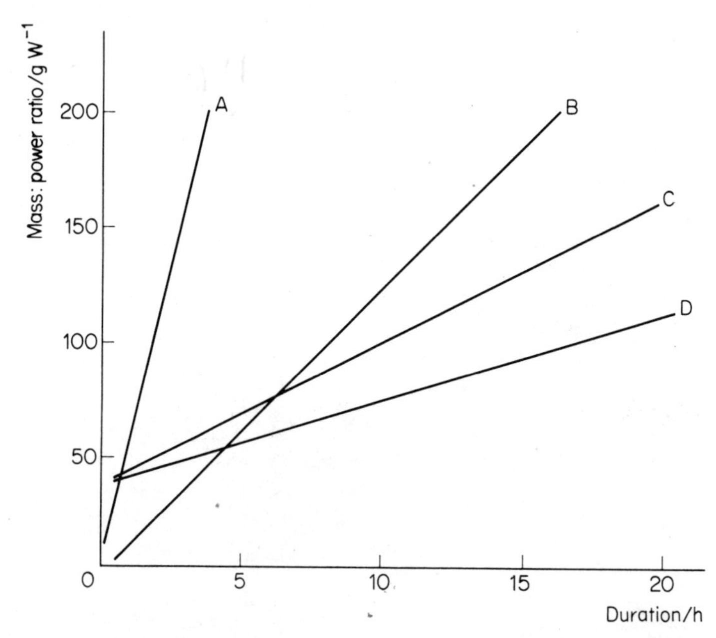

Figure 13.1 Mass : power ratios for various kinds of complete storage battery operating over different periods of time. A = lead–acid battery; B = silver–zinc battery; C = hydrogen–oxygen fuel battery, with gases stored in cylinders; and D = hydrogen–oxygen fuel battery, with gases stored as liquids at low temperature

important, and here the low temperature hydrogen–oxygen cell is likely to show an advantage over conventional secondary batteries. For power production over periods of more than a few hours, the mass of the secondary cell is greater than that of the fuel battery together with the stored gases. This is illustrated graphically in figure 13.1. However the problem associated with such low temperature cells has been discussed extensively already in earlier chapters: specially prepared fuels such as hydrogen, carbon monoxide, hydrazine or methanol have to be used because of the difficulty in obtaining materials which are electrocatalytically active at these low temperatures, and the preparation or manufacture of these adds considerably to costs. It is fair to point out a further advantage of these fuel cells, however, in their easy 'recharging' (by replacing the gas supply cylinders) compared with the long times required for ordinary secondary cells.

During the remainder of this chapter we will consider some areas of possible employment of fuel cells which seem particularly favourable, and then look at some future developments which might affect our economic arguments of chapter 12.

Transport

The possible use of fuel cells to power light industrial trucks has been mentioned previously, and an illustration of a vehicle of this type is shown in figure 13.2. This flat truck was constructed and used in a pilot experiment

Figure 13.2 Lansing Bagnall tow truck fitted with $3\frac{1}{2}$ kW hydrogen–oxygen fuel cell battery. It can run for a full 8 h shift between refuelling, and had $3\frac{1}{2}$ years' use at Chloride laboratories

by Chloride Technical Ltd some years ago, but the general development was not pursued, presumably for the usual economic reasons.

More recently, attention has been turned to the possibility of fume free and silent vehicles for urban transport; it is generally accepted that the motor car we use today cannot continue as a normal mode of transport in the cities of the future. It is too large, too inefficient and, some say, too much of a hazard to our environment to be very suitable in circumstances where most people who can afford to do so will use personal in preference to public transport. A solution that has been offered is the provision of small, light, low powered electric cars (which could be privately or publicly owned) simply for city travel. Such vehicles seem well suited to being powered by fuel battery systems, and some consideration was given to this in chapters 11 and 12.

The amount of space required by conventional cars, both when driven and when parked, is also a source of worry for city planners who generally realise the greater efficiency (in that sense) of rail transport, while accepting the loss of personal freedom and convenience that this mode of urban transport involves. In an attempt to try to reconcile these apparently contradictory requirements, an interesting scheme of 'dual mode' urban vehicles has been proposed and described. These vehicles would be capable of running on electrified railway tracks provided on main routes and of independent operation (on more or less conventional roads) in the congested central urban areas where everyone's journey is likely to follow a different route, or in the outer residential areas where similar conditions may prevail.

Clearly this arrangement could combine the merits of high density rail transport on main routes with the flexibility of private car operation off these main routes. However a prerequisite will be a compact electric power supply for operation away from the electrified tracks. The supply could be ordinary storage batteries or storage batteries of an advanced type like the lithium–chlorine or sodium–sulphur system or fuel batteries. The vehicle would be driven by a dc motor receiving power either from an electric conductor rail or wire or from the battery. There are other interesting details of this idea but they do not concern the use of fuel cells in such vehicles; fuel cells offer some advantage over other battery systems for the reasons outlined in this section.

Other forms of transport powered by fuel cells seem less likely to be developed in the future. The General Motors 'Electrovan', made in 1967 to demonstrate the feasibility of a fuel battery operated vehicle of this type, contains cells of estimated capital cost of about £3000 kW^{-1} whereas a price competitive with the internal combustion engine would have been about £7 kW^{-1}. This enormous difference is partly due to the use of platinum as an electrocatalyst—although some of this would be recoverable at the end of the life of production vehicles—and it does therefore illustrate the dependence of future applications of fuel cells on the development of cheap electrode materials that can be used with readily available fuel.

General power supply

The various characteristics and advantages of both low power and high power fuel cell supply systems have been extensively discussed in chapters 11 and 12, and the place that might be occupied by fuel batteries can easily be seen. It appears more likely that interest will develop in the smaller scale power units for emergency and stand by use or, probably more significantly, for district power supplies in areas remote from conventional generating stations. As we have seen already, electricity transmission costs are very high and it may well be advantageous to use local supplies of natural gas or even to pipe it over considerable distances, since this is a very much cheaper process. A ratio of between 6 : 1 and 18 : 1 for cost of electricity transmission compared with transmission of natural gas has been quoted, the figures allowing for the efficiency of conversion. One scheme of this sort envisages a separate power plant for each domestic or small industrial user whereas what seems possibly to be a more practicable idea is to site 1 MW fuel battery power plants in places such that they can supply about fifty houses. The supply would be low voltage dc (50 V has been suggested). This kind of supply would avoid very large numbers of cells having to be operated in series and would also remove the costs of supplying and operating an inverter to produce ac. It seems likely that most household appliances could operate satisfactorily on dc and in some cases (for example, lighting) low voltages might be advantageous. The overload capacity of fuel cell systems mentioned earlier would have considerable attraction in this kind of scheme.

Fuel cells and the hydrogen economy

The arrangement, described in the previous section, of 'satellite' low power fuel battery electricity supply plants fuelled by a pipeline network of natural gas could fit in very satisfactorily with the ideas of a fuel economy based on hydrogen. This concept is discussed more fully in a companion book in this series, but we can see here that if piped hydrogen became readily available then the small scale fuel cell system is even more attractive, since a more satisfactory low temperature cell can be used. Some measure of regeneration might also be included. We have seen in chapter 11 that fuel cells can be used as secondary batteries in certain circumstances, forming hydrogen and oxygen by electrolysis in the 'charging' mode. Thus an increased peak hour or overload capacity could be provided by this means, underground storage tanks of hydrogen and oxygen (produced by the electrolysis) being brought into use. The hydrogen–oxygen cell operating at low temperatures has instant starting characteristics and can be made quite sensitive to load.

Power for electrolytic processes and electrosynthesis

There are some commercial chemical and metallurgical processes which require large currents of dc electricity; for example metallic aluminium is produced industrially by electrolysis of the ore (bauxite, Al_2O_3) dissolved in molten cryolite (Na_3AlF_6). It has been suggested that a possible future development might be to produce the power required for such processes from a battery of fuel cells. It seems odd, perhaps, to convert chemical energy into electrical energy and then promptly convert it back again, but there could be advantages, particularly since a direct current supply is essential.

A slight modification of this idea is to use the hydrogen produced when chlorine is manufactured by electrolysis of sea water or brine to power a fuel cell which might provide some of the energy necessary for the electrolysis. The hydrogen is, in any case, a byproduct of this process. No doubt there are other examples, and one worthy of mention is certainly *electrosynthesis*, particularly of the starting materials for polymer manufacture. For example electrolysis of a concentrated aqueous solution of a tetralkylammonium alkyl or aryl sulphonate containing acrylonitrile produces adiponitrile, an important precursor in nylon manufacture:

$$2CH_2{=}CH{\cdot}CN + 2e^- + 2H_2O \rightarrow NC{\cdot}CH_2{-}CH_2CH_2{-}CH_2{\cdot}CN + 2OH^-$$

Future improvements in fuel cells

It is unlikely that we can predict what sort of improvements will take place in the design and operation of fuel cells in the near future but we can certainly look at the places where improvement is needed. Undoubtedly much work is desirable on electrocatalysis generally and particularly to look for what would probably be the most significant breakthrough possible in fuel cell technology—a catalyst suitable as electrode material for a low temperature cell using hydrocarbons directly as a fuel. Unfortunately there seems to be no clue as to how such a material could be obtained. Even the electrocatalysts known to be useful for hydrocarbon fuels at higher temperatures and for hydrogen low temperature cells have unattractive features. In particular, the metals concerned (nickel and platinum, generally) are not very abundant and it has been calculated that the world's nickel production would be severely strained by any large scale construction of fuel cells. In fact it was estimated that about 4.5 kg of nickel would be required per kW of power. Moreover, a rather more alarming result was obtained from a calculation about resources necessary for constructing platinum electrodes; it has been suggested that the world's total production of platinum would only be sufficient to supply about 1 per cent of our electrical power demand.

The question of fuel production and supply is considered in the next section but there are other problems too. Constructional difficulties are not severe but the availability of materials suitable for gasket and insulator production is not as good as it might be. Many high and medium temperature cells eventually fail because of damage (by corrosion or otherwise) to these components rather than because of any shortcomings in what we might regard as the essential items like electrodes and electrolyte.

Electrolytes at present seem to be the most satisfactory part of the fuel cell. The only two features at all worrying are the rather limited choice of materials available for the higher temperatures of operation, and the poor performance of the oxygen electrode in acid electrolytes, which has been referred to several times.

A considerable improvement in power: mass ratios for hydrogen–oxygen cells would be possible if the gases could be safely stored in devices other than the very heavy steel cylinders generally used. It is true that in principle both hydrogen and oxygen could be stored as liquids, thus occupying less space, but the necessary thermal insulation and ancillary cryogenic apparatus might well increase the total mass to a figure greater than the conventional high pressure cylinders. This consideration would, of course, be most significant for fuel cell powered vehicles or other portable power supplies.

Choice of fuels in the future

Many of the important considerations governing choice of fuels have already been discussed elsewhere in this book, and we need here only to look at possible changes the future may bring. The economic changes in the supply of oil since 1973 are likely to have two effects: first there will be increased enthusiasm for any device (such as a fuel cell) which appears to convert the energy of oil to electrical power at an efficiency better than that currently obtained with conventional systems and, secondly, interest may shift from oil back to coal as a basic source of chemical energy. Thus the fuels produced by the gasification of coal (probably hydrogen, methane and carbon monoxide, although methanol is also a possibility) may become increasingly important. Moreover, particularly in Britain, natural gas will become highly significant as a major energy source even if only the most pessimistic prophets of North Sea discoveries are to be believed.

A further source of hydrocarbon fuel likely to be tapped in the near future is the gas escaping from oil wells when the oil reaches the surface. At present (mid-1975) this gas is usually burnt as waste at the wellhead, but there are serious proposals for converting it by a reforming process to methanol. The prospect of cheap methanol might well change the economics of methanol cell operation enough to make it a worthwhile competitor for the diesel engine, although at the time of writing the outlook is not hopeful.

Fuel cell operation always requires the provision of 'clean' fuel, which usually means free of sulphur containing compounds which are likely to poison any electrocatalyst used. Fortunately this is not too much of a problem, particularly if the fuel is produced by any reforming process, since the feedstock for the reforming plant has to be sulphur free too in order to protect the catalysts also used in this process.

Final comments

We have seen quite clearly that the theoretical basis for preferring direct electrochemical conversion of energy by fuel cells to the currently used indirect methods is well founded. We have also seen that a very great deal of effort and money has been spent in investigating the scientific and technological consequences of this theoretical advantage. In fact, the technological development of the fuel cell has progressed a very long way in a very short time: very satisfactory fuel batteries have been constructed with high efficiencies and long working lives and their characteristics have been demonstrated for a wide variety of applications. Unfortunately, economic calculations have always shown that the fuel cell is not quite good enough to supersede conventional systems where commercial considerations are overwhelmingly significant, as they must be in most circumstances.

The possibility of further technological advance is slight, and also (unfortunately perhaps) is likely to be accompanied by technological improvements in the more conventional varieties of energy conversion, thus making the competition even more critical. For example, there is a good chance of the gas turbine and the Wankel (rotating piston) internal combustion engine becoming much more significant in the fields of power generation and transport respectively.

We must also not lose sight of the fact that a very considerable investment has been made by all civilised countries in the manufacture of items connected with our present conventional forms of energy conversion. This, of course, applies most particularly to the enormously important motor car industry and its associated activities. Balanced against that perhaps we should remember that the skills required for fuel cell construction are not likely to be great and the machines needed will not have to be very sophisticated, but on the whole it seems possible that a change in our pattern of life may be a necessary accompaniment to the widespread adoption of fuel cell systems.

It is probable, therefore, that any substantial change of emphasis in energy production towards fuel cells in the future will depend not so much on further technological advance but more on shifts in the more purely economic and political climate, which may favour the efficient, clean and quiet fuel cell.

BIBLIOGRAPHY

Thermodynamics

J. R. W. Warn, *Concise Chemical Thermodynamics*, Van Nostrand & Reinhold (1969)

P. A. H. Wyatt, *Energy and Entropy in Chemistry*, Macmillan (1967)

E. B. Smith, *Basic Chemical Thermodynamics*, Oxford (1973)

Reaction Kinetics

P. G. Ashmore, *Principles of Reaction Kinetics*, Chemical Society Monographs for Teachers, No. 9, 2nd edn. (1967)

Electrochemistry in general, including electrochemical kinetics

E. C. Potter, *Electrochemistry, Principles and Application*, Cleaver-Hume (1961)

J. O' M. Bockris and A. K. N. Reddy, *Modern Electrochemistry*, MacDonald (1970)

J. O' M. Bockris and D. Drazic, *Electrochemical Science*, Taylor & Francis (1972)

Electrolyte Solutions

R. A. Robinson and R. H. Stokes, *Electrolyte Solutions*, Butterworths, 2nd edn. (1970)

Fuel Cells

A. B. Hart and G. J. Womack, *Fuel Cells*, Chapman Hall (1967)

K. R. Williams, *An Introduction to Fuel Cells*, Elsevier (1965)

Air depolarised Cells

D. P. Gregory, *Metal-air Batteries*, Mills & Boon (1972)

Industrial Chemistry

P. Wiseman, *An Introduction to Industrial Organic Chemistry*, Applied Science (1972)

SI Units

M. L. McGlashan, *Physiochemical Quantities and Units*, Chemical Society Monographs for Teachers, No. 15, 2nd edn. (1971)

Electrochemical Data

R. Parsons, *Handbook of Electrochemical Constants*, Butterworths (1959)

Fuel Cell Economics

J. Verstraete et al. in *Handbook of Fuel Cell Technology*, (C. Berger, Editor), Prentice-Hall (1968)

INDEX